Lean Evolution

Lean thinking is a powerful method that allows organisations to improve the productivity, efficiency and quality of their products or services. Achieving these benefits requires good teamwork, clear communication, intelligent use of resources and a commitment to continuous improvement. This book shows how lean thinking can be applied in practice, highlighting the key challenges and pitfalls.

The authors, based at a leading centre for lean enterprise research, begin with an overview of the theory of lean thinking. They then explain the core tools and techniques and show how they can be applied successfully. The detailed implementation of lean thinking is illustrated by several case studies, from a range of industries, in which the authors had unprecedented access to the management teams.

With its focus on implementation and practical solutions, this book will appeal to managers at all levels, as well as to business students and researchers in lean thinking.

Nick Rich is RCUK Innovative Manufacturing Fellow and Director of the Innovative Manufacturing Research Centre, Cardiff University.

Nicola Bateman is a senior research fellow at the Lean Enterprise Research Centre, Cardiff University.

Ann Esain is a senior research fellow at the Lean Enterprise Research Centre, Cardiff University.

Lynn Massey is a RCCPI Research Fellow at Griffith University Queensland Australia.

Donna Samuel is a senior research associate at Lean Enterprise Research Centre Cardiff Business School, Cardiff University.

Lean Evolution

Lessons from the Workplace

Nick Rich
Nicola Bateman
Ann Esain
Lynn Massey
Donna Samuel

CAMBRIDGE
UNIVERSITY PRESS

University Printing House, Cambridge CB2 8BS, United Kingdom

Cambridge University Press is part of the University of Cambridge.

It furthers the University's mission by disseminating knowledge in the pursuit of education, learning and research at the highest international levels of excellence.

www.cambridge.org
Information on this title: www.cambridge.org/9780521843447

First published 2006
First paperback edition 2012

A catalogue record for this publication is available from the British Library

ISBN 978-0-521-84344-7 Hardback
ISBN 978-1-107-40719-0 Paperback

Contents

List of boxes *page* vi
List of figures vii
List of tables ix
Glossary x

1 Introduction 1

2 Understanding the lean journey 11

3 Understanding your organisation 32

4 Laying the foundation stone of CANDO 60

5 Visual management and performance measurement 80

6 Problem solving, TQM and Six Sigma 95

7 Pull systems 122

8 Total productive manufacturing (TPM) 141

9 Sustainability 163

10 Group learning 185

11 Reflections and future challenges 200

References 207
Index 210

Boxes

2.1	Pressures for change	*page* 12
2.2	The five principles of lean thinking	15
2.3	Toyota seven wastes	17
2.4	The management activity list	24
3.1	Buying into lean	45
3.2	Establishing a clear message	46
3.3	The change model	53
4.1	Cosmetics and CANDO	62
4.2	Clean up at Health Products	65
4.3	Health Products and arranging the workplace	66
4.4	Neatness stage at Health Products	66
4.5	Promoting champions	68
4.6	Maintenance repair and overhaul facility	71
4.7	Involving operators and the consensus approach	72
4.8	Medical Consumables and the MD's office	73
4.9	Mornington Cereals	75
4.10	Mornington Cereals (too much talking and not enough action)	76
4.11	Views from the Learn 2 companies	78
5.1	Too many measures! What's the most important?	90
5.2	The daily walk	91
6.1	Air Repair mapping	97
6.2	Medical Devices case	100
6.3	Steel corporation case	113
6.4	Medical Devices case	117
6.5	Air Repair journey of improvement	118
7.1	High variety: low volume pull systems	136
8.1	TPM definitions	146
8.2	Calculating OEE	151
8.3	Single point lesson (SPL)	157
9.1	Definitions	164
9.2	Visual representation of lean implementation	168

Figures

1.1	Knowledge flows within the Learn 2 network	*page* 4
2.1	The house of lean	26
2.2	Lean improvement stages	29
3.1	Who's doing what?	48
3.2	Value stream mapping (adapted from Rother and Shook, 1998)	51
3.3	What customers want passes horizontally through departments	56
3.4	The overall production system features	58
4.1	The lean model	61
4.2	CANDO linkages	62
4.3	CANDO red tag (Rich, 1999)	64
4.4	On going improvement measurement	69
4.5	The Steering Committee structure	70
5.1	Role of visual management	81
5.2	Placement square	83
5.3	A typical communications board	87
5.4	Data trail	88
5.5	Graph with target	92
5.6	Visual management linkages	93
6.1	Quality filter map for Air Repair 2000	98
6.2	Problem solving at the different levels of an organisation	99
6.3	Problem resolution summary board	103
6.4	Problem solving is integral to the achievement of improvement and change	104
6.5	Quality links in the lean journey	106
6.6	The seven basic tools of quality (Bicheno and Catherwood, 2004)	108
6.7	Ishikawa diagram (also known as a fishbone or cause and effect diagram)	108
6.8	Converting a problem into a statistic to provide a practical solution	111
6.9	Primary source of variation	112
6.10	New philosophy of quality	113
6.11	Measure systems analysis	115

6.12 ANOVA results for a gauge R&R review (output from Minitab) 116
6.13 Prioritisation matrix for final testing 117
6.14 A multi-vari chart is a tool identified for use in the seven basic tools of quality (output from Minitab) 118
6.15 Combination of quality and lean activities to improve the cost of quality 120
6.16 Supply chain development and integration 120
7.1 The house of lean 123
7.2 The lean wheel 123
7.3 The sequence of kanban 126
7.4 Kanban levelling box (heijunka) 128
7.5 Kanban flow 129
7.6 Simple kanban floor squares 130
7.7 Signal kanban (slow cycle control) 132
7.8 Alternative kanban signals 133
8.1 House of lean 142
8.2 The lean wheel 142
8.3 Asset criticality 158
9.1 Structure of lean implementation 164
9.2 Sustainability of PI activities 165
9.3 Rolling out from PI activities to strategic level 167
9.4 Enablers for Class A and B activities 169
9.5 PDCA for shopfloor ideas 170
9.6 Enablers for Class A activities 172
9.7 Frequency graph of enablers 174
9.8 Frequency graph for inhibitors 175
9.9 Frequency chart of engaging issues 177
9.10 Frequency chart for resolution of engaging issues 178
9.11 Resources required for full roll out of PI activities 180
9.12 Sustainability links to other chapters 183
10.1 Primary knowledge flows within the Learn 2 research programme 187

Tables

3.1	Design logic of the traditional and lean firm	*page* 34
3.2	Traditional and lean supply chain	35
3.3	Lessons from lean programme successes	37
3.4	Waste and appropriate maps	50
4.1	CANDO and its various forms	63
8.1	The Toyota seven wastes and TPM	148
8.2	Involvement and roles	150
8.3	Eight major pillars	153
9.1	Processes to incorporate enablers	181
10.1	MOPS and Learn 2	192
10.2	Champions and agents understanding of lean	193
10.3	Benefits	198

Glossary

ABC	A means of categorising products, failures or other group of observed issues such that the most important sources can be identified in terms of the impact and volume. 'A' classifications are therefore the most important and 'C' the least and this allows problem solving to be directed to those issues/products with the most potential benefit to the company.
agile manufacturing	The ability to accommodate change responsively in terms of volume and mix flexibility.
Andon	A subset of visual control management which is used to signal abnormalities with the production process or to identify deviations between the desired pace of the work station/assembly line and takt time requirements.
autonomous maintenance	Those activities of routine equipment maintenance conducted by individuals and small groups to a level of safety and quality assurance established by the business/engineering specialists. This is the front line of maintenance activity and is used to detect and correct abnormalities quickly.
CANDO	Also known as the 5S system or workplace discipline and control
cellular manufacturing	A layout choice which involves the co-located configuration of machinery in a manner that the output of one machine directly feeds the next or feeds a small buffer before the next. The ideal cell adopts an approach of 'one piece flow'.
changeover	The time taken from the last good piece produced from the existing batch of work at a machine to the first 'accepted' good product from the new batch. A concept developed by Shigeo Shingo.
constraint	The bottleneck or limiting factor (either equipment, human or management policy) which limits the throughput and output of production. A concept developed by Eli Goldratt using the 'theory of constraints' approach to operations management.

continuous improvement (kaizen)	From the Japanese meaning 'virtuous circle'. Meaning small step changes in performance as a result of continued analysis and process changes to improve the efficiency and effectiveness of production or administrative activities.
control item	An element of a product or production system used to assess whether the system is working within an agreed specification. Used to prompt action upon detection of variation which may cause defects or instability.
defects	The manifestation of an error within the production system which results in 'un-saleable' products or stopped administrative process. An error represents a deviation, by humans or machines.
error proofing	The design of processes and devices which prevent the creation of errors and defects through physical means, i.e. the prevention of accidental consumption of tablets through the introduction of caps to medicine bottles that can only be operated by adults (push, twist, and turn) or the use of the physical size of the product to prevent misalignment (i.e. a 3.5 inch floppy disk can only be entered into a 3.5 inch drive in one exact way).
Just-In-Time	The generic name given to the logistics of the Toyota production system as opposed to the lean enterprise which covers the production, management and supply chain processes.
kaikaku	From the Japanese meaning radical break to the circle of improvement. This approach is a very condensed and intense activity conducted within the factory to make an instant improvement in performance and to demonstrate that change can be instantaneous.
kanban	From the Japanese meaning 'ticket' and referring to the information cards used to trigger the removal and manufacturing of replenishments within a manufacturing system. The cards cycle between internal customers and suppliers to ensure production occurs Just-In-Time.
lead time	The total time a piece of material resides in the production system from start of production to finished goods. Also used as quoted 'order receipt to delivery time' when interacting with customers and this includes all administrative processes, manufacturing, queuing and despatch activities.
LERC level scheduling	The Lean Enterprise Research Centre, Cardiff Bussiness School. The lean approach to smoothing production requirements over a time period so that the same amount is produced every week etc. The logic is not to batch production and incur long periods between making the same product, but to cycle quickly through the entire range of products so as to minimise delays and limit any queues.
MRP	Material requirements planning. A computerised scheduling system, originated in the 1950s, which served to time the arrival of materials from

	suppliers and from within the factory through computer-calculated and printed schedules. These systems did not calculate whether capacity was available to produce.
MRPII	Manufacturing resources planning. An extension of MRP which utilised advancement in computing power of the 1970s and also more sophisticated algorithms to find the 'best fit' schedule that matched customer delivery dates and available capacity to produce.
NVA	Non-value adding or actions conducted by the organisation that adds no value to the product and serves to increase costs. Activities for which the customer would prefer not to pay.
OEE	Overall equipment effectiveness – the baseline measure of all TPM activities and a means of charting progress (through trend analysis) of improvement activities.
one piece flow	The smallest manufacturing batch size and unit of flow around a cell. One piece is taken from raw to finished stage in one single loop of activity involving 'walking' operators that handle multiple machines to manufacture the product. It may then be processed further by another operation.
Policy deployment	The process of setting a 3–5 year business goal and annual improvement challenges to all business functions. The total population of middle managers agree the key projects which will enhance the competitiveness of the firm and collaborate to ensure their execution. The hidden lesson of lean manufacturing.
product family	A group of products that share a similar value stream, product characteristics and/or sales patterns. These stock-keeping units are grouped to form a band of products that are used to create volume for a cell design or a means of analysing the critical flows of materials within a factory.
pull	The approach to triggering manufacturing operations using the 'kanban' replenishment system, whereby movements of finished products trigger re-supply from internal processes.
push	The generic term used to define manufacturers that schedule production and often operate 'make to finished stock' operations systems.
right first time	A surrogate measure for zero defects which applies to perfecting production and administrative processes such that the activity never involves an error.
single minute exchange of dies	The term applied to changeover activities of less than ten minutes. This is the base line of quick changeover improvements and will lead, through re-engineering, to 'one touch' exchange of dies (OTED). This process is

	important as it allows smaller batch sizes and more variety of production (scope) to be achieved in a single shift.
Six Sigma	A powerful new approach to quality management, which represents a goal (target value) and a methodology. It fully supports lean production.
standardised work	The codified and visual documentation, written by operators and specialists, to assist learning and conduct of repetitive operations. Displayed at the point of use – these documents are the basis for improvement activities as much as they are the standards that sustain a common way of working.
takt time	The rate of production needed to equal the average rate of product sales to customers. This calculation is used to ensure flow processes are performing effectively or that buffers are being replenished at the desired rate.
target cost	An approach to determining product features and costs using a backward process of deducting marginal, distribution costs to derive the maximum cost of production. This cost is used as a challenge to product and process designers. The objective is to meet the cost or design products with lower costs to enhance margins and profitability.
TPM	An approach and methodology for creating 'robust' production systems through improving the technical skills of the workforce and the specifications of the manufacturing technologies employed.
TPS	Toyota production system – the logic and implemented features of the originator of the lean system.
U cell	The preferred layout of a lean cell due to the ability to reduce and flex the amount of labour needed through 'walking single piece flow' (also known as motion kaizen) by the operator.
value adding	An activity for which the customer is prepared to pay – typically a transformation process within the value stream.
value stream	The internal activities which must come together to produce an output and more broadly those processes within each tier of the supply chain which span raw materials production to consumer. Includes order fulfillment and design value streams.
value stream mapping	A portfolio of techniques used to visualise and diagnose the current status and future potential improvements within and beyond the factory. Essentially management techniques for operations system design purposes.
visual control	An approach to visualising the status of a process and to make deviation in performance readily identifiable without need for specialist training.

	Examples include vehicle dashboards and warning lights which prompt the driver to stop etc.
work in process	Also called 'work in progress' and contains all materials held within a factory that are part finished and lie between raw and final stages. Approach is to standardise and minimise this level of materials until production flow can be used to displace the inventory.
zero defects (ZD)	The ultimate goal of all companies that seek to compete on quality of product. ZD is a measure of internal process control as much as it represents fault-free customer service to external organisations/consumers.

1 Introduction

Donna Samuel

I'd like to do some improvement work, but I'm always too busy.

If this sounds familiar to you, then you are the person for whom this book has been written. Continually displacing what is important for what is urgent is endemic amongst western managers. We call it the 'fire-fighting syndrome', because we have an image of a manager with a fire hose who spends his time fending off the problems of the day. The trouble is, of course, if we never make time for what we know to be important, we may not have a business tomorrow!

This book addresses the messy subject of lean implementation. If you are involved in business improvement and have figured out that it is one thing to find out what to do but another to actually do it, then this book has been written with you in mind.

Let us address the meaning of lean first of all. If you are familiar with the car industry or even the manufacturing sector more widely, then you may have come across the term. The term 'lean', coined by a group of academics, concerns the ability of the Toyota Motor Corporation to achieve outstanding manufacturing performance levels in Japan. They wanted a word to capture what they saw – a system without 'fat'. By fat we really mean waste and we'll come back to that later on. Toyota's manufacturing operations are impressive in many ways, but the distinguishing characteristic that these guys noted was that Toyota was able to do a lot more with a lot less. In other words, with a lot less resources (inputs), they are able to produce a lot more (outputs). Their productivity (measured in terms of headcount and unit output) was *double* that of equivalent firms in the west. And that productivity performance was not at the expense of quality either. In fact, their quality performance (however you cared to measure it) was dramatically higher than their western counterparts (in the order of 20 times). So the phrase *lean production* was born. The academics involved documented their findings in a book, *The Machine that Changed the World*, which was published in the early 1990s. The book was hugely popular and, in 1992, was voted business book of the year. Over a decade on, many industrialists are still keen to take the 'lean' ideas and use them within their own organisations. For lots of them, understanding these ideas is simple enough (many are about getting the company back to basics and good old commonsense). However, translating and applying the ideas is a different matter. Knowing what to do is the easy bit, doing it is rather harder. This book is about doing it.

The main purpose of this opening chapter is to capture your attention and make sure that, having read this introduction, you are inspired to read on. This introduction, then, aims to inform you of why this book has been written and for whom it is intended. By the time you have finished, you will have decided whether it is worth your while reading further or not.

Who should read this book?

This book is targeted mainly for business improvement managers or managers with some other similar titles. Improvement may be every manager's dream – to be able to step back from the day-to-day drudgery of operations and dream of slicker and more streamlined processes sounds fantastic. Improvement can, however, be a double-edged sword. Depending on the culture prevalent within the organisation, being handed the job of improvement can be a poisoned chalice. However, business improvement for most presents a unique and gratifying challenge.

After all, every company is trying to improve itself in some way or another. Some are caught in the 'fire-fighter syndrome' trap; others have seen dramatic improvements already and want to know where to go next. Wherever you are on the improvement road, if you are involved in the manufacturing sector, this book has been written with you in mind. If you happen to be involved in car manufacturing, much of the terminology and many of the concepts that will be addressed in this book may be familiar. Nevertheless, you will still find this book to be something of a departure from many of the others and here's why.

Our unique selling proposition (USP), as we like to call it in business, or the thing that differentiates this from other books available on the market, is that it tackles improvement implementation. Implementation is messy and cloudy and, for that reason, most authors leave it well alone or make reference to it only in passing. Consequently, amongst the current improvement/lean literature, the practitioner really only has two types of book available to him: the management guru texts (such as *The Machine or Lean Thinking*) or toolkit books (such as *The Lean Toolbox*). The guru texts are a good overview, but lack the detail you would need to apply the concepts. The toolbox books are great reference documents, but give the impression that familiarity with an array of tools and techniques is the only armour you need to transform your business. In our opinion, what is missing amongst the lean literature is a text that has been written with the user, the business improvement manager, in mind. This book aims to address that gap.

So how do we address the issue of lean implementation?

If you have decided by now that you do roughly fit within the target audience, you will be eager to know how we intend to tackle the difficult issue of lean implementation. Essentially this book has been produced as the end result of a three-year research programme. The research programme is entitled Learn 2 – a simple play on words. All the companies involved were all *learning* to (Learn 2) be *lean*. The next section offers the reader some background and context to the programme.

The Learn 2 programme

The programme was an idea that sprang from a group of researchers working at the Lean Enterprise Research Centre (LERC). LERC is a research centre based at Cardiff Business School. Details about LERC and what goes on there can be obtained by looking at our website.[1] The centre was built upon the work of Professor Dan Jones, co-author of *The Machine that Changed the World* (1990), the text within which the term 'lean' was coined. The Centre carries out leading-edge research into the application of the lean concept in different environments (particularly outside the automotive sector, the original home of lean). LERC research staff have considerable cumulative knowledge about the implementation of the lean concept but have never formally collaborated on the issue of lean implementation. Consequently, the Learn 2 programme, as devised by Dr Nick Rich,[2] was a significant departure from the traditional research programmes conducted within LERC.

Learn 2 is a small network of manufacturing companies, from a range of different sectors. They were joined together in a network for no other reason than they were all engaged in making their organisations lean or leaner and were seeking help with lean implementation.

The programme was launched in May 1999. Prospective companies were gathered together and told that their needs would best be served by participation in the Learn 2 network. In order to take part, companies would have to pay with a nominal fee to fund the university research programme and this fee provided them with an amount of 'support capacity' from designated LERC 'mentoring' research staff. The details of how they used that capacity would be worked out with each company individually, based on their needs, wants, histories, etc. The programme would enable them to tap into the cumulative implementation knowledge that resides within LERC and, as such, this latent knowledge represents the main benefit companies would receive

[1] www.cardiff.ac.uk/carbs/lerc

[2] Dr Nick Rich is now a Director of Cardiff University's EPSRC Innovative Manufacturing Research Centre.

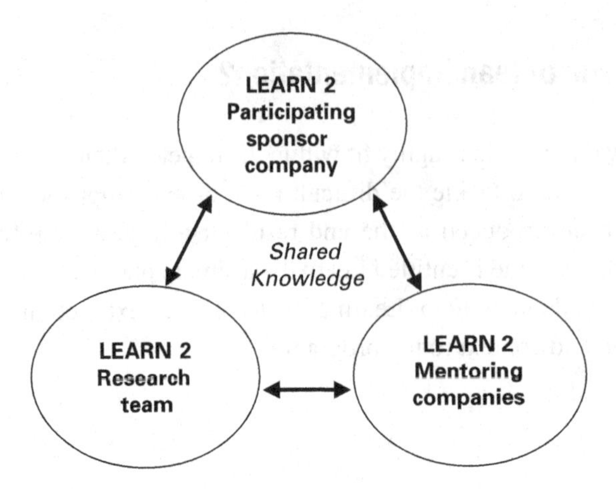

Figure 1.1 Knowledge flows within the Learn 2 network

from participation in the programme. It is important to note that, although companies were signed up in spirit for the duration of the programme (i.e. three years), they were entitled to leave earlier. If, for example, after the first year they felt they were not getting good 'value for money' by their participation in the programme, they could withdraw.

So, access to LERC expertise is the main benefit participating companies would receive, but not the only one. Learn 2 companies were encouraged to participate in the 'networking' aspect of the programme as well. Each company assigned someone to perform the role of main contact with LERC and to act as internal programme coordinator. We refer to these individuals as lean change agents. They would meet on a quarterly basis to exchange ideas and share information on progress. Other representatives from participating companies might be invited along to the meetings, but the core individuals would remain unchanged from meeting to meeting. The 'networking' aspect of the Learn 2 programme has proved to be very powerful for a variety of reasons and is further explored later in the book.

From LERC's perspective, the value of Learn 2 lay in the interchange of ideas among those members of research staff who were assigned to the project (the LERC Learn 2 research team). The programme provided a unique opportunity to study a collection of disparate organisations, united only by their common objective to implement lean. The Learn 2 network therefore offered the LERC Learn 2 team a chance to observe and reflect on what seems to work well and why.

The Learn 2 network is wider than the participating companies and, in total, includes: the LERC Learn 2 research team, the Learn 2 participating companies and three mentor companies. These mentor companies are ones with whom LERC has had an ongoing relationship for many years and who have themselves been through the turmoil of lean implementation. Their participation in the programme is informal and they are used to provide support and encouragement. The main information and knowledge flows within the network are shown in figure 1.1.

The Learn 2 participating companies

For reasons of confidentiality we have not revealed the identity of the actual companies that took part in the programme. Our view is that any company trying to improve itself should be commended; after all change is never easy. However, improvement always and inherently involves criticism of what went on before. That is the nature of the beast. In our experience, there are few commercial organisations that like to 'wash their dirty linen in public' and, in order not to compromise our findings, we have decided to 'anonymise' the participating companies and to disguise them in other ways. A brief profile of participating companies follows.

Heavy Products

Heavy Products is a manufacturing site, part of an international group, engaged in the production of large metal-based products. The company employs over 500 'long-serving' people. The motivation to begin the journey to lean implementation was stimulated by increasing competition from overseas manufacturers and falling market prices for this 'commodity' type of product. The manufacturing facility may be classified as 'low volume and high variety' with an increasing specialisation on the finishing of products which can be sold as 'higher value added'. It is probably fair to say that this company was facing a serious crisis.

Cosmetic

Cosmetic is a manufacturer of high fashion products, many with short lifecycles, for retail businesses and has a huge and growing product range. The business employs around 550 full-time employees and an additional contingent of temporary labour to meet their highly seasonal demand. The business, recently acquired and added to an international group, was motivated to engage in lean production as a means of improving the performance of the firm and to generate the profits necessary to fund the expansion of the group. The culture of this highly unionised factory is a healthy bias towards learning and involvement of the workforce, but is also characterised by cautious apprehension at the management level in applying 'lean' to such a 'high-variety' environment. The business works to many different customer standards and the factory has a high priority on 'good manufacturing practices' (GMP) to ensure the process of conversion is controlled to the highest standards.

Health Products

The growth rate of Health Products has been phenomenal, based upon high customer demand, but the market for these products has attracted a number of competitors from countries with low-cost labour. The resultant pressure for 'cost performance' has highlighted the importance of factory productivity and the senior management of the business, heavily committed to 'world class' manufacturing and growth, have selected the lean approach as most appropriate for their business needs. The process of conversion is complex and highly technical, engaging almost 2,500 workers at multiple sites, to service a common warehouse, which, in turn, services the world market. The business is dominated by sales to Japan, the most demanding product market in the world, and is regulated by the Food and Drugs Administration.

Air Repair

Air Repair is a long-established business, making and repairing complicated and critical systems for aerospace and aeronautical customers. Employees at the site are very experienced, unionised and technically skilled in working with high-precision and computerised manufacturing technology. The business is an independent manufacturer, supplying final assemblies of the finished product and again is classified as a 'high-variety and low-volume' business, which is heavily regulated by external awarding bodies. The business also has a complicated supply chain and was motivated to 'lean' by the need to control complex product flows and to improve its customer service by compressing lead-times and reducing arrears.

Medical Devices

Medical Devices' products has grown from a cottage-style industry of laboratory operations to large-scale production. This transition has caused a number of problems for the highly educated workforce, and automation has brought with it major issues concerning product and process control. The business, owned by a multinational corporation, was experiencing major operational difficulties, causing a great deal of stress and tension within the business. The business employs around 400 employees to produce a growing range of product families, in high volumes, to customers located throughout the world. The site is heavily regulated by the Food and Drugs Administration.

Mornington Cereals

Mornington Cereals make breakfast cereals and are a profitable organisation within a highly successful group. At the start of their lean journey, however, the company were experiencing considerable competitive pressures from within their own group, in particular the European sister companies, and they were also recovering from the unsettling effects of shutting down part of the factory and having to reorganise staffing levels. They decided to embark on their lean journey in order to combat these competitive pressures and consolidate new staffing structures. Mornington Cereals manufacture breakfast cereal from a range of wet and dry ingredients. The product is made through a process of mixing, extrusion, drying and packing. The final packaged product is made available to the consumer in a range of forms from single portion pouches (30g) to 1.5 Kg boxes.

It is clear that the Learn 2 network embraces a highly diverse range of companies, each of which has its own story, motivations and prevailing conditions shaping the nature of their lean implementation. However, the commitment and enthusiasm of each company, primarily at the outset, was pervasive.

The mentor companies participating in the programme included:

1 A major retailer who has turned the fortunes around in the last 40 years or so and who now leads the way within their own, highly competitive market.
2 A retailer of electrical and electronic components who has cornered a significant proportion of their market through their unique customer service offering.
3 A large-volume automotive company that has been forced to find innovative ways to survive the last few decades in this highly volatile and fiercely competitive sector.

In summary then, and in answer to the question posed at the start of this section – how do we address the difficult and messy issue of lean implementation – we adopt a 'systems' approach. In our observations of all companies participating in the programme, we take an holistic view. We frequently ask: does the system or subsystem (whether it be the order fulfilment system or the new product development system) work? Is it predictable and can abnormalities be easily detected and addressed? These questions constitute the essential ingredients for the success of Toyota's production system, the archetypal lean manufacturer. Our conclusions, however, are in no way prescriptive. We begin from the premise that lean implementation will always be contingent upon the set of circumstances impinging on the organisation at any given time.

How this book has been structured

A brief explanation of how this book has been structured is worthy of mention at this point to save the reader potentially wasting valuable time. As mentioned before, the book has been written with the industrial practitioner in mind. For that reason we have tried to give a cohesive logic to the structure, without making the chapters overly interdependent. In other words, the book can and really should be read from beginning to end to produce a cumulative message; however, at the same time, each chapter can be read on a stand-alone basis as well. Therefore, if you are interested in visual management or preventative maintenance, but do not have the time to read the entire text, then you can simply dip into the relevant chapter and use the book as a reference text. The undesirable by-product of designing the book in this way is inevitably a certain amount of repetition. We have endeavoured to keep this to a minimum and also to signpost and cross-reference as much as possible.

The book has been broken into four main parts:

The first part includes Chapters 1–3. Collectively these chapters provide the reader with a certain amount of contextual knowledge, thereby sensitising the reader for what follows in later parts. Chapter 2 is entitled 'Understanding the lean journey' and explains the background, rationale and logic of lean. While we do have a model to describe lean implementation, the detail of lean implementation is as varied as the companies that took part in the Learn 2 programme. This chapter explains why this is the case and, in doing so, explains why a prescriptive approach to lean implementation would be both inappropriate and misguided. Whilst this text inevitably offers advice on what to do and why, it is never intended to be dogmatic and dictatorial. Management intuition should not be undervalued in determining what is and what is not likely to work in any given organisation. We *always* take notice of it when we are facilitating companies. The final chapter of this part, entitled 'Understanding your organisation' encourages the reader to carefully consider the set of conditions surrounding his or her own particular organisation by comparing them with stereotypes. The issues discussed in this chapter should help the reader focus his or her thoughts on a number of important aspects, such as the relative likelihood of success or failure of lean implementation within their organisation, the most appropriate starting place and the likely pace of change. In the first part, therefore, we guide the reader from the general to the more specific in preparation for the more detailed discussion in the parts that follow.

Chapters 4 and 5 sit together and form the likely starting place for many lean implementation initiatives. Indeed most, though not all, of the Learn 2 companies started here with varying degrees of success and difficulty. The reason for the popularity of CANDO (see Chapter 4), leading to visual management and performance measures (chapter 5) as the focus for the first year of a lean implementation initiative, is not that this is always the right starting place, rather that, for many organisations, lean is the 'baby' of a handful of enlightened individuals within the organisation and therefore company-wide buy-in is often a dominant problem and force for inertia. Although, we do not advocate a prescriptive solution for lean implementation, CANDO and visual management can be particularly powerful ways of achieving this buy-in and of mobilising the enthusiasm and energy of others, many of whom will become important actors in the later stages of lean implementation. The benefits of CANDO and visual management (VM) are precisely that they are visual and that they have a direct and positive impact on those that matter most, the people who touch the product. Management love it too because their factory is transformed, in a very short space of time, into something much more inviting for customers to visit. Operators love it because it makes their working lives and conditions better and easier and they instantly perceive lean as something that involves them. Consequently these two chapters are full of the learning points that were pulled together as a result of the collective experiences of the companies participating in the programme. Many improvement initiatives begin with a whole raft of information-finding activities, leading to lots of meetings but little actual change. In lean we are always careful to avoid 'paralysis by analysis'!

In the fourth part, chapters 6–8 together form the crux of lean and address what we refer to as the three pillars of lean: total quality management, the Toyota production system (pull and just-in-time) and total preventative maintenance. Our position is that any company seeking to become truly lean will have to master all three pillars (as Toyota has done), though not necessarily all at the same time or in the same way. Our Learn 2 companies embarked on a three-year programme to initiate an organisational transformation. Toyota, for example, began the transformation process over half a century ago and is still engaged in perfecting the system. Also, it would be unrealistic and unhelpful to limit our discussion in these chapters to the Learn 2 companies alone. Consequently to illustrate and enhance our discussion, we draw on our wider knowledge and experience of lean implementation.

In the final section, which includes chapters 9–11, we largely come back to the Learn 2 companies as the primary sources of our deliberations. In chapter 9 we address the important issue of sustainability. How to maintain and evolve the change programme is a real issue for many organisations who may have started their programme some time ago. Here we explore the issue by explaining some of our own research into this issue. The research findings culminate into a series of enablers and inhibitors which, when practitioners are made aware of them, can help to shape and improve their own chances of success. In chapter 10, entitled 'group learning', we deliberate on some of our own lessons as a result of the Learn 2 programme. Many of us were struck by the power of collective learning. The feeling of support that comes from being part of a like-minded group is not dissimilar to being part of a family. We believe, though we cannot prove, that one or two of the participating companies may have abandoned their lean journey during certain points of crisis if it had not been for this support. Finally in this part, we attempt to synthesise our thoughts in a closing chapter entitled 'Conclusions and future challenges'. Here we weave together many of the issues raised in the preceeding parts of the book. We reflect on where we are, where we have come from and begin to consider issues that will be important for the future.

Finally, it should be noted that each chapter includes a diagram designed to guide the reader to other relevant and related information. Also, each chapter ends with signposts and recommended readings. Here we suggest a selection of what we believe to be the very best from the raft of literature that is out there.

That completes the primary purpose of this opening chapter. Hopefully you have read enough to want to read on. At the outset it is important to remember that improvement is about dynamic change and therefore those involved will inevitably experience periods of highs and lows. Within the Learn 2 programme, we saw the initial excitement and flurry of activity give way to lethargy and disenchantment in the second year. We hope that this book offers practitioners a guiding light to remind them that, although the journey is not an easy one, where we are going to is far superior to where we have come from. The third year witnessed the true integration of management with the lean initiative and the strategic integration of all business processes rather than just those close to the point of production. The problems of the second year resulted from a natural point in the improvement process, where the production teams cannot progress further because they are constrained by traditional practices and inhibiting behaviours in the management and support departments. Year three is the point where management begin to set direction and reduce demarcation between operations and support functions.

2 Understanding the lean journey

Nick Rich

Introduction

The market for manufactured products is getting more and more competitive. There is no such thing now as a safe product or a safe market. Even companies that previously had a monopoly position, such as telecommunications, gas or electricity corporations, have found themselves deregulated or the target of consumer groups. These changes add up to the new competitive world within which consumers and customers have greater power and increasingly demand higher levels of customer service and greater value (Womack and Jones, 1996). To compound these changes, Western manufacturers face a new threat 'from the East', as India, China and former soviet bloc countries, each with massive reserves and large numbers of manufacturing firms, increasingly look to sell products into Europe. These 'exports' drive up the levels of competition and therefore no Western firm can afford to stand still and not to become involved in factory-wide improvement activities. Failing to understand the new competitive environment is one way to lose market share and another is not to improve to protect customers and keep them loyal, as other businesses compete to trade with them.

These competitive conditions have generated a new 'set of rules' for Western manufacturers. These new rules include the provision of the highest level of customer service, the delivery of quality products in shorter and shorter lead times and product proliferation to offer variety to customers (McCarthy and Rich, 2004). If you take a few minutes to consider what life was like ten years ago and compare it with the present, your business has probably moved from recording percentage defects produced during manufacturing to 'parts per million' (PPM) levels, it probably offers more products than ten years ago and has probably halved its lead times. These are the new rules of competition and they are there to eliminate all waste from the business and to ensure the customer gets real value from trading with your firm.

These new conditions are far removed from those of the past. Then, there was a belief that, if a business was successful, it would continue to be so. This belief is now very questionable and the present is no indicator of the future. There was also a belief that product patents would protect a manufacturer from competition, but this too must be questioned as it is not uncommon for competitors to 'strip down' your products and find ways of making them cheaper, and even better. Another belief was that buying the latest technology would provide a means of defence against competition but, as productive assets are available on a world-wide scale, this offers no true and sustainable advantage. So there is no such thing as 'safe' market, no product or technology that is immune to these increases in competition.

Box 2.1 Pressures for change

- New and emerging manufacturing economies with low labour costs are attracted to mature Western markets where they can exploit their 'cost advantage'.
- The power of the internet in purchasing materials and components on a global scale and therefore access to alternative suppliers has increased exponentially. As such, power has shifted to the customer/consumer.
- Deregulation of European markets has resulted from international trading agreements and this has liberated trade and increased competition for manufacturers.
- Uncertainty about the national economy and policy decisions concerning greater integration with Europe to further lower trading barriers and intensify competition.

You may be forgiven for thinking that, with all these pressures, manufacturers face inevitable decline, but this is not so. Instead, the art of modern management is to harness the intellectual and creative capabilities of all the workforce and target these capabilities to produce better products more cheaply, thus achieving the 'world class' manufacturing that sets your business apart from its competitors. Once all is said and done, the future of a manufacturing business depends upon its employees. People are the source of new ideas, new ways of working and they determine the long-term success of any improvement initiative. Whilst some businesses like to put up posters stating that 'Employees are our number one asset', those companies that understand their markets and value the customers who consume their products will really believe in developing the skills of their managers, their team leaders and the different teams who come together to produce the quality products that provide value and satisfaction to customers every day of the year. To these companies, working life involves improvement and 'working together'. Competition is not a concept that generates fear for these businesses because it is treated as just 'the way things are'. These businesses are, to quote a popular expression, 'on a mission' to be world class and inflict the pains of 'catch up' on their competitors. This spirit is, to some observers, difficult to understand, but it unites the companies that have been involved in the research upon which this book has been based. It should also be pointed out that, although this book will explain the changes and processes involved, it will take you a number of years to master the practices that these companies have already adopted. Further, you may replicate the practices, but you cannot bank on having the quality of employees at these firms. What unites these businesses is that they are well down a never-ending route to achieve 'world class' performance based upon an approach known as 'lean thinking'.

What is lean thinking?

From the 1980s, many models of 'how to run' an efficient and effective manufacturing organisation were developed. Some of these models were promoted by academics and others by consultants. Each was used to convince business managers to change what they currently had and replace it with a new approach to business. However, some of these models were 'flights of fancy' and others were conceptually sound but did not offer any logic or process for implementation, other than ripping everything up and starting again. Few companies have the luxury to simply find a new site and start all over, and for years managers, academics and consultants struggled with how to get change and improvement from established 'brown field' operations. These facilities had existing technology (in various states of decay); they had employees with experience, but also with customary practices; they had suppliers, who also remembered the past; and they had managers, who found it difficult to begin a change process that could last many years. These issues are not small and could include some pertinant questions. How do you convince a work force that is suspicious of change to 'buy into' a vision of a new manufacturing facility within which they would be expected to contribute ideas? How do you persuade every manager to change their behaviour so that they achieve the most from this new approach? And, how do you tackle these and still get the product out of the door? Without answers to these fundamental questions, many of these new models, attractive alternatives that they were, lacked a real logic and toolbox of techniques for managers to begin the transformation process.

It is no surprise that later 'high-performance' manufacturing models (being termed 'world class manufacturing') were influenced greatly, during the late 1980s, by the rise of Japan as a manufacturing nation, and also as prestigious Japanese corporations began to establish manufacturing and assembly facilities in North America and Europe. These companies included Toyota, Nissan, Honda, Sony, Panasonic and a legion of other Japanese firms that few people have heard of but were exemplar companies in their chosen product sector. This rapid rise, from nowhere, to become one of the world's centres for manufacturing excellence set in place a quest to find out how these companies designed and operated their manufacturing systems to achieve such competitive advantage. This process was to yield a new and post-mass production model of manufacturing that has been termed 'lean production' and more recently 'lean thinking'. Despite the origins of the lean approach being the Japanese car industry, it has been emulated widely by Western manufacturing businesses engaged in a wide variety of market sectors from contact lens production to helicopter systems overhaul.

The origins of the 'lean approach' can be traced to US fears that the newly emerging Japanese vehicle assemblers held a competitive advantage over their established Western counterparts. These fears prompted benchmarking studies of the global automotive industry to test these fears and to find the causes of any such advantage. The results of these studies are reported in the publication, *The Machine that Changed the World* by Womack, Jones and Roos (1990). For Western manufacturers, this text provided the first data, drawing from the automotive industry, that Japanese manufacturers enjoyed a 2:1 productivity and a 100:1 quality advantage over the West. These gaps were huge and clearly showed that high speed was attributable to excellent levels of quality performance by the Japanese manufacturers and all its suppliers. Further, the authors found this form of manufacturing used less inventory, less space, less effort and less capital to produce. Hence the term 'lean' was coined to describe this buffer-less approach to manufacturing.

The term 'lean' also contrasts with the stereotype of mass production (the dominant form of manufacturing in the West), with its images of large batch sizes, high costs of inventory and vast spaces occupied by stock that was slow moving or worse dormant/obsolete. These features of the mass production system all include slow-moving materials and constant interruptions between receiving a customer's order, fulfilling that order and getting paid. The most advanced form of lean production, and the company that had designed this new way of manufacturing, was Toyota. As such, not all Japanese car companies were the same, and Toyota was a truly global exemplar in assembling high-quality vehicles at very high speed. The study also found that, with the Toyota–General Motors joint venture in the USA, the lean approach had been successfully imported to an established 'brown field' site and had resulted in superior performance that nearly matched the levels achieved by established Toyota sites in Japan. These improvements had been achieved at a site with existing workers, existing practices and a Western culture; but, more importantly, the finding suggested that, with careful attention to the design and implementation of change management programmes, the 'lean' approach could work, and generate improved performance, in Western manufacturing organisations (Rich, 1999).

From this point onwards, the automotive industry has emulated the lean approach of Toyota, but also other industries began to take an interest in the 'lean approach' for their own market sector. In this manner, Western businesses began to experiment and implement lean methods and to make improvements. The deliberate 'importation' of these practices by Western companies (and non-automotive businesses) attracted a new study that resulted in the publication *Lean Thinking* by Womack and Jones in 1996 (two co-authors of *The Machine that Changed the World*). This new publication followed the progress of 52 Western cases, as they redesigned their existing operations to become 'lean producers', and describes the 'before' and 'after' benefits achieved by these organisations that ranged from a manufacturer of shrink wrapping machines to highly complex aerospace component manufacturing. These studies not only proved that 'lean production' could be transferred with competitive benefits for Western businesses, but dispelled a general belief that 'lean production' was culturally specific to Japan. The generic lessons of the imported lean approach was summarised by the authors into five principles of 'lean thinking'.

Box 2.2 The five principles of lean thinking

1 **Understand value** in terms of '**WHAT**' the customer wants to buy and what provides customer satisfaction/customer service. This principle highlights the vital role of giving customers products they value and are prepared to pay for and promotes value enhancement through 'waste' elimination. Waste includes all those activities, in the current production system, that stop or delay the process of converting materials/information into customer payments. At a second level, understanding value, also includes the design of products that enhance customer satisfaction profitably and this is the art of 'making the right product in the right way'.
2 **Identify the value stream** and the internal activities undertaken within the firm that convert a customer order into a fulfilled order and the activities associated with generating new products for customers (**HOW**). Once you understand how you manufacture and design products, you can improve the process and from here you can begin to work with the wider value stream (suppliers and customers) to eliminate all the wastes between companies involved with satisfying customers.
3 **Make products flow** is the third pillar of lean thinking and involves keeping materials and information moving so that materials 'flow' to customers without delay or interruption. Stocking materials for very long periods of time reduces stock turns and this inflates costs and ties up huge amounts of capital in materials that are not being sold for a profit.
4 **Pull production at the rate of consumption** is used when it is not possible to complete flow products to customers (due to the number of customers, short lead times, the needs of your technology and batch sizes or other constraint). Under this rule, where it is not possible to flow production, a buffer must be deliberately designed to allow customer orders to be fulfilled from a carefully managed stock point. In this way, it is possible to maintain customer service by later production and finishing processing, 'pulling' out the work they need to complete orders from this buffer point. For advanced forms of lean production it is possible to have many small buffer points that are used to directly link internal customer and

supplier production operations and allow customer orders (removed from finished goods stocks) to completely pull work through the factory. This is known as the kanban system at Toyota and allows instant availability of products and short lead times simultaneously.

5 **Seek perfection** in every aspect of the business and its relations with customers and suppliers is the final pillar and rule of lean thinking. Here, the authors stress the use of problem-solving teams of operators, managers and inter-company teams to squeeze out the last remaining elements of waste and non-value added activity.

These five simple principles seem, at first sight, very logical and rational, but this logic is counter-intuitive to traditional Western thought. In the West, big batch sizes have always been regarded as the best way of making products cheaply, as it keeps unit costs down, but this can be at the expense of flexibility, lead to unsold stock and does not allow for improvement or progress. It can be no real surprise that the leading Japanese 'lean producers' are also businesses that have invested much problem solving time in getting set-ups and machine changeover times down to an absolute minimum. In this way, you can change machines over quickly and the production system can cope with high variety. Also they have been at the forefront of Total Quality Management (TQM) and have practised it since the 1960s to eliminate the wastes of defects and the sources of production system problems (Shingo, 1986).

At Toyota there is a set list of 'wastes' to be eliminated – known as 'the seven wastes' (Ohno, 1988). For Toyota, the value of production occurs only when supplied materials are being converted into a finished form for sale. This means that the actual production cycle of the asset is valuable (so long as it is producing good product) and that any other activity does not add value but instead adds a cost (as someone is paid to move products around the factory etc.). Seeing every activity in the workplace in these terms will make you very frustrated as you now begin to see the difference between very fast cycle times and the very slow nature of processes that add cost but no value to the product. In this way you are thinking like a 'lean customer' – you are looking for the wastes that don't give you any value and for which you would rather not pay.

When you have reached this level of 'lean awareness' your frustration with the 'current state' of the factory can get unbearable, but all you have achieved is a personal revolution – to create a factory revolution you will need to convince others that there is 'waste' all around them too. This is not an easy task. You are telling employees, who come to work to do a good job, that a lot of what they do is 'waste'. This requires careful handling and the realisation that it is not the fault of the individual, it is just a failing of the current production system. After all, operators do not tend to determine batch sizes, nor are they responsible for the factory layout – these decisions were forced upon them by others. In convincing others that waste exists within the factory, help is at hand, and for over 50 years Toyota have worked with a fixed list of seven key wastes and these can be used to allow your co-workers to identify aspects of workplace management that are not as they could be.

Box 2.3 Toyota seven wastes

1. The waste of **OVER-PRODUCTION** where vast amounts of products are made in batches and simply 'dumped' into finished goods or work-in-process. This result when there is a mismatch between customer demand for products and the ability of the production system to make to that demand. This is one of the greatest problems with mass production and the reliance upon large batch sizes.
2. The waste of **UNNECESSARY INVENTORY** where the results of over-production and other 'unimproved' constraints means that inventory is simply held awaiting an order in the belief that future orders will come.
3. **INAPPROPRIATE PROCESSING** is another waste that results from a mismatch between the processes needed to make a product and the processes that are in place. In this manner, many firms use very sophisticated machinery to manufacture simple products that would be best produced using 'simpler' and less-expensive technology. Typically, in the West, large sophisticated machines with high processing speeds tend to be 'pumped full' with production in order to ensure a 'pay back' for the asset, and keeping such machines occupied with work inflates batch sizes and generates inventory (two forms of waste).
4. **UNNECESSARY TRANSPORTATION.** A further form of waste concerns the movement of materials around a factory from the receiving bay to the shipping bay. This activity can consume many hours and involve many kilometres of transport (with each activity offering the potential for product damage).
5. **UNNECESSARY DELAY** concerns the simple 'dwelling' time as products are ready to be converted, but sit waiting. For much of factory time, materials will be 'idly hanging around' in an uncontrolled manner.
6. **UNNECESSARY DEFECTS** is the production of materials (that consumes value adding time) but have to be reworked or scrapped. In this way valuable capacity is lost forever – you cannot reclaim it even by working overtime. So imagine the problem of large batch sizes, long travel distances and, hidden within these batches, defective products!
7. **UNNECESSARY MOTION** occurs when the production process is poorly designed and operators engage in stressful activities to handle materials. This is an unusual waste (ergonomics), but, as claims for repetitive strain injury rise, many firms are facing large settlement fees from employee claims and solicitor bills.

You will also find that these seven areas are well known to factory employees – who have probably given up moaning about the problems they cause. But it is important that the knowledge of what is value and what is waste is communicated and that, for any lean change programme to work, this knowledge forms the bedrock of all improvement activities. So the starting point for all future activities is the clear differentiation between 'value', aspects of what the company does, for which the customer is prepared to pay, and those aspects that are waste or costs and should be reduced or eliminated altogether. This is principle number one in the understanding of and differentiating between value and waste.

Armed with this knowledge, lean thinking promotes an understanding of what factory activities are undertaken to fulfil orders or to bring new products to the market. These activities within the factory form the value stream and involve a series of hand-offs between departments as information is converted into requests to manufacture and then to ship to the customer. Each of these elements of the manufacturing process contains waste and unnecessary costs and each can be improved, often using very simple tools and techniques. Further, once the internal value stream has been stabilised and improved, it is then possible to regulate suppliers and customers so as to improve the total production chain for raw materials, suppliers, your factory and its customers (even its customers' customers). Unsurprisingly, this level of integration involves removing inter-company wastes as well as improving supplier performance by equipping them with the ability to detect 'waste' from 'value'.

These five basic lean principles, each involving a substantial amount of time and effort to 'master' represent the basic elements of the lean system. These common elements underpin every implementation project.

Your lean journey

The 'lean journey' of change management and improvement is however a journey that is unique to each company. Some companies will be motivated by poor trading conditions and low margins and some will be riding the 'crest of a wave', with massive business growth and increasing market share. It does not matter what motivates the drive to implement 'lean thinking', it is the careful design of the implementation plan (and its logic) that is important to building a sustainable factory system within which all employees offer suggestions about how to improve. Also there are no 'magic wands' and simply copying what others have done is not a business strategy. The 'lean' techniques some businesses seek to copy were answers, engaged by another firm, to meet its own business problems. The art of lean production is to identify correctly the problems within the business and its material flow processes and then to select the appropriate techniques to solve these problems. These responsibilities belong to the factory management. Therefore the plans and structures that support the implementation process are management duties too. In the past, this has not always been the case and resulted in a fragmented and poorly thought-through approach to change (involving convincing operator teams that change was legitimate). The consequence was that most management demands for improvement were greeted with a belief that it was just another fashion that would soon pass, fail or quickly be replaced by another fad. The credibility of managers amongst shop floor teams under these circumstances was understandably low. Further, most managers never 'walked the talk' but demanded change and quickly reversed decisions or disbanded implementation teams with the first signs of improvement (if any). If you work for one of these types of firms, it is probable that you will also have to market the lean improvement programme to fellow managers before approaching the shop floor teams. If you don't 'market', then life on the production line may change and improve slightly, but, without the support of managers (who control key business processes and decisions such as the size of production batches), improvements will not be dramatic and your job will still have high levels of waste and frustration.

The five 'lean thinking principles' provide a systematic way of looking at the business and aligning all activities with the generation of value-added for customers and profits for the firm (they appeal to managers mainly). The pillars also provide a framework with which to engage shop floor personnel and therefore unite, on application, the elimination of waste with the elimination of business weaknesses. Lean thinking should therefore not be seen as a box of techniques that can be applied indiscriminately in the hope that improvement will result, lean thinking has a distinct logic and a process of improvement where, at each stage, you must select the appropriate tools to solve your own problems. In short, its no magic silver bullet, but the lean journey is a continual process of questioning 'why' the business works in the way it does and finding new ways of adding value to what customers wish to buy. In this manner, the application of 'lean thinking' is unique to each business, to its problems and its relationship with its customers and suppliers.

Misconception and myths

There are many misconceptions that relate to the lean approach and undertaking this approach to changing the design and operation of the firm implies certain choices. These misconceptions are rarely discussed in books about 'lean' and instead many authors 'gloss over' or choose to ignore them. The misconceptions include:

- **'Lean production is just a series of tools and techniques'** – This is a misguided view of lean implementation and is unfortunate. It is held by people who have not understood the basic logic of the approach and individuals who love the visible elements of imposing change in the factory (the techniques employed). This view misses one vital element – the role of management. Others might say that lean production is focused only on operator teams and this is partly true in that lean organisations will seek to stabilise the physical product flow before engaging in new management processes that create business capabilities (i.e. design for manufacture, time to market, environmental management etc.). Creating operational stability is a precursor to stopping fire-fighting and allowing managers to return to managing.

- **'Lean production only works for Japanese companies'** – This is a common point of view and usually precedes, 'But we are NOT Japanese.' Well, it does work for Japanese companies, this is true (just look at their market shares or benchmarking data), but are the Japanese so different as human beings that they are superior to Western workers? Surely not! Western workers are more than capable of engaging in lean production. This view is merely a means of stopping any form of experimentation and improvement in the factory environment. Many of those who utter this statement tend to hold fixed and underlying views about the workforce too (and these views do not rate workers very highly either).
- **'Lean is stressful for workers' and 'lean is mean'** – The amount of stress that workers face is a reflection of the design of any system (a management responsibility) and is not typically associated with a lean working environment. The lean approach is to get the best out of people – not to flog them to death. The latter was the mass production and scientific management approach which was reinforced by 'piece rate working'. Lean businesses are careful to develop and integrate their core workforce and to promote co-destiny of worker and business. Much of the HR systems of lean companies is devoted to the promotion of 'harmony' and 'joint interest' in managing the business. As you will see during the remainder of this book, most lean methods lower stress and improve morale, and need workers to be involved in the change process. That is not to say that the lean way is a utopia but that, typically workers are not 'flogged' or stressed to the point that they leave (a waste of training and development investment made by the company in problem solving and technical skills). Further, if you are lucky enough to visit a company that is engaged in lean production, cellular working, housekeeping and problem-solving groups, ask these workers whether they would go back to their old ways of working. You will find that few will answer 'yes' and wish to return to the old days of being treated as a pair of hands that are unattached to a brain. But no matter what we write here, you can only judge for yourself by asking and studying some of the cases we will present in this book. For senior managers, the lean approach is based upon business growth and support for the flexible deployment of the core workforce throughout their working lives. It is silly to believe that lean companies would treat their workers, who have enjoyed high levels of lean awareness and training investments, as a variable cost that is simply employed and unemployed at will. Lean companies recognise their investments in people and therefore, to increase the competitive capability of the firm, seek new markets and new types of products to make. Such an approach is one of growth, not the meanness associated with 'de-layering', 'downsizing' and losing workers with highly valuable and transferable skill sets.

- **'Lean applies only to the car industry'** – This is a common citation at the beginning of a lean change programme. This is an interesting response and implies that the individual understands some of the features of the lean approach, but cannot make the connection between what has developed in the car industry and what could be taken and used effectively in your factory. A simple trawl of the internet reveals a wide variety of firms (operating in a diverse range of sectors). Just look at the promotional materials you get each week heralding the next lean conference and look at the companies presenting before you reach for the bin – you will find most have no connection with cars at all. And none of the cases, we will provide in this book, has any association with making cars or components for them.
- **'Lean cannot work in a unionised environment'** – the old chestnut! This is a real falsehood and again says more about management attitude than reality. Many lean organisations work closely with trade unions and are proud of their partnerships. In many respects, a management team gets the union it deserves or has inherited and vice versa. Managers are dependent upon their trade union and implementation processes are far easier when the union is actively engaged in the change process. For small businesses it is also true that many trade unions have training programmes that can be used and are relatively inexpensive (covering issues from health and safety management to advanced forms of problem solving). So having a trade union has few barriers and many advantages if the management–union relationship is managed correctly. You will also find that the Japanese lean manufacturers in the UK tend to work with single union agreements covering the entire workforce (why do this if you hate trade unions?).
- **'We have tried it before and it has failed so why bother doing it again'** – This is perhaps a sad reflection on industry's ability to exploit good innovations and hold on to the benefits. It is unfortunately a legacy of 'past' management regimes and their ability to maintain improvements. The statement also suggests that, as managers, we have historically failed to truly understand the improvement technique and how best to transfer it into daily usage. The simple truth is that a lot of the tools and techniques associated with lean production are known; they are also quite straightforward and they have been proven to work at other firms. So really this retort is a criticism of the management at the factory and possibly the absence of management during change processes. Workers do not tend to deliberately 'kill off' good initiatives that help the business to grow and develop. As such, this criticism is not one solely of lean production, but of management in general. The answer is that lean businesses spend a great deal of time communicating the need for change and allowing (rather than imposing) the techniques needed to be selected by the workers themselves. Lean businesses also support these initiatives and review the progress of teams frequently, including recognising effort as well as the achievements of teams. Communicating the need for change and making the link between change and business needs are therefore important features of starting and sustaining the improvement process.

Most misconceptions are quite funny and you should revisit them as you near the close of your first year of implementation and change. They are, however, common concerns that people have or have picked up – most importantly they are reasons to 'hide behind' and not to change so they need to be worked through in advance. You might as well address them before you announce and launch your lean programme as this will help give it credibility. It is better to get them 'out in the open' and let people voice their concerns. At the end of the day, even the sceptics will sit back and let you get on with it (let you convince them that you are telling the truth and this is not another management whim). You must also recognise that change in any form is threatening to some employees, and success/sustainability means you should spend time (even on a one-to-one level) with the key people upon whom you will rely in getting changes made.

Before we begin to explore the features of the lean approach it is worth pausing to consider the implications of what we have just been through, not least because without understanding, the implementation process will be severely slowed down.

- Any organisation considering the implementation of the lean approach must understand that the business is a single 'system', effectively making every element of a business dependent upon others. Think about it. If I am a maintenance technician, I can only do my job if production control allow me time and access to the machinery. If I am in production control, I cannot guarantee quality output if maintenance technicians are prevented from conducting their routine tasks.
- A lean manufacturing business will always seek 'growth', as, without growth, continuous improvement will lead to job losses. This is a major lesson, and should have been written in a huge typeface. All lean exemplar companies are committed to full employment of their core workforce.[1] These people are often called 'associates', for the simple reason that they may work anywhere within the growing factory over their lifetime of training and employment. So I could start my working life in production control and then move to aftermarket parts, or I could transfer within my business from 'line A' producing vacuum cleaners to 'line D' producing washing machines. These movements are important to spread knowledge. A misconception of the lean approach is that it leads to lean human resources (i.e. less people). That is not the case but it may mean you protect your core workers and use temporary labour to meet high production targets. Nonetheless, lean companies protect and nurture their core workers – it is these people upon whose creativity that competitive advantage rests!
- **Simplify everything to show instantly any sign of abnormality** is another tenet of the lean approach. In a production system with low buffers, it is important that any small sign of deviation or abnormality is detected quickly and that, at the point of

[1] A core workforce comprises of the full-time employees at the factory as opposed to any temporary labour that is used as a buffer to supplement core workers in meeting production targets.

detection, people have the skills to quickly solve and eliminate the problem. So lean organisations manage improvements from a basis of stability.

- **Co-operation is better than confrontation.** Another overriding belief of lean organisations is that co-operation is the key to all improvement. Therefore managers – whom we have said are dependent upon each other – should co-operate to get done the things they cannot do as individuals. Further, internal production teams should co-operate to improve and this thinking is extended to a trusted supply base upon whom the business is dependent for its material inputs.

These basic elements of the lean DNA are important and worthy of discussion by managers. There are many misconceptions as we have seen, so getting to grips with the basic underlying assumptions of 'going lean' is important. It should be remembered that, if any single senior manager states openly or behaves against these principles, then this will be obvious to the workforce and exploited by those who will gain by 'sinking' the lean or any other initiative. Having got this far, we will now take a look at the common elements – those that can be seen – of a lean system.

The core of the lean production model

Whilst every lean journey is unique, there are certain features of the model that are common to all lean implementation models, regardless of whether you are making petrol engines, converting nuclear fuel rods, or rolling aluminium foil. These features can be broken into two key areas. The first area concerns management and the second area concerns the change process and operational personnel (usually in teams). It is important to clarify these two areas because in Western industry many senior managers (and directors) also want to play with the change process (in the same way they often become high-paid 'progress chasers').

The correct definition of the management role is that of the 'system planner and designer' – that is the true value of management and what shop floor teams look to managers to provide. Senior managers and directors should engage in planning activities and enjoy the feedback sessions from teams who have implemented change within their business. Middle managers (those who head up business departments) are the true system designers and the most difficult group of managers to engage cross-functionally to think about 'processes' that generate 'value'. Middle management therefore lead change (leaving the senior managers to champion programmes) and will provide support to operational improvement teams. To be effective, change programmes must focus on re-aligning the current production system, finding problems to be solved and thinking, as a group of managers, about the future redesign needed to enhance value, and working cross-functionally to improve relations and performance between business functions. We will explore some more of these activities in the next chapter, but engaging middle managers is vital to the sustainability of improvements and to the amount of savings which result. However engaging managers is not enough – the business needs a mechanism through which techniques can be deployed by the teams and the teams supported as they engage in change. This responsibility falls to the 'promotion office' (Womack and Jones, 1996).

The 'promotion office' is a group of employees who are assigned to teach, facilitate and monitor the change programme. They are provided by management as a sign of commitment and as a resource that is utilised by the improvement effort on a factory-wide scale. For some companies, this involves secondment, and, for others, current employees work in this part of the organisation, but generally these individuals are skilled (from a technical and a lean perspective) to drive forward improvements and quantify the savings made by the teams.

Box 2.4 The management activity list

- Planning the change programme, setting out factory policies, providing resources, reviewing progress and measuring performance.
- Membership of cross-functional management improvement groups to map and change the systems that support the workplace.
- Mentoring, key reports and developing their technical knowledge.

For many books written during the early 1990s, the management role was presented as one of being a 'coach' or 'facilitator' and that is the nature of a lean change programme. This was a complete 'U turn' from the 1980s that promoted the de-layering of business and resulted in eliminating managers altogether. But by the 1990s, the theme was definitely that of the manager as coach. As a manager you cannot and do not usually have the time to lead change directly. Further, by leading change and dictating what happens, none of your reportees ever mature into the role as leader and consequently every worker still looks to the manager to make all the decisions. So by interfering unnecessarily you are generating waste and stopping progress. You must also not be too quick to 'get stuck in' to correct situations where progress is slow or you know the answer to a problem – instead you should remain silent and typically the team will, after an embarrassing pause, return to provide you with alternative solutions. This is the hardest of all lessons for managers – it is not a weakness, it is a clever way to get your reports to get back to their value in finding ways to improve the factory or at least their part of it. Further it is a great way to test the 'learning' within the organisation and tends to result in employees finding the answers to problems or innovating and inventing solutions for themselves.

For managers to be successful as team coaches, they must also develop teams. Such teams cannot be regarded simply as operators clubbing together for a bit of problem solving, or worse 'socialising' under the pretence of problem solving. So let us define some more value related to organisational roles.

Within any manufacturing organisation there are middle managers and these people can add value because they tend to know what is 'best practice' in their area or function and also where to find it (or at least tell you which companies are good examples). So these people can add value by designing new ways of working, helping teams and continuing to learn about 'best practice'. Yes this also means visiting other companies with an open mind to discovering ways of innovating and spotting common problems.

Another set of employees who can add value can be found in the specialists of the firm. These specialists come in all shapes and forms and can be found in maintenance engineering roles, purchasing, accounting, product design, industrial engineering and many other departments. What makes these people 'valuable', even though they may often not be managers, comes from the skills these people have acquired through apprenticeships, training at technical colleges, education at universities and also by membership of professional bodies. These bodies include organisations such as the Chartered Institute of Purchasing and Supply, the Institute of Logistics, the Institute of Mechanical Engineers and many more – too many to list here. These people have specialist diagnostic skills and also have access to 'best practice' through these experiences. The best way these people can add value, therefore, comes from engaging them in problem solving and project work – not conducting reactive or purely administrative procedures (for which they do not use their intellects or do so in a very individualistic manner!). These people are great for problem solving and can, in a single decision, save businesses millions and radically improve business performance. Areas where they can excel include how to improve the quality of the total business, how to improve delivery performance, how to eliminate and compress time between customers' orders and delivery, how to get products to market in a quicker and less problematic manner and also how best to logically reduce the operational costs of the firm. They have not traditionally been considered people who should join teams, especially as members of cross-functional management teams looking at business issues, but to ignore these individuals who often are desperate to take part in change programmes is a sin in the modern era when we expect everyone to help the business survive and grow.

Finally, we have the traditional focus of team working – that of operator teams. These people, by far the greatest number of employees, make good products, keep improving processes and maintain the standards of the factory. Lean companies put a lot of time and effort into these employees, nurturing their skills and keeping material flowing.

So the 'manager as a coach' approach to modern manufacturing management is correct and can only be exercised when we combine correctly the talents of organisational employees and focus them on key business-related issues to improve the firm. Just simply demanding the introduction of techniques is not lean – it looks like lean but is really a superficial veneer. Lean is about eliminating weaknesses within the business and preparing it to compete for customer business in the future when customers and competitors will have changed their quality expectations, delivery expectations, cost expectations and product range expectations of the future market place. Any business

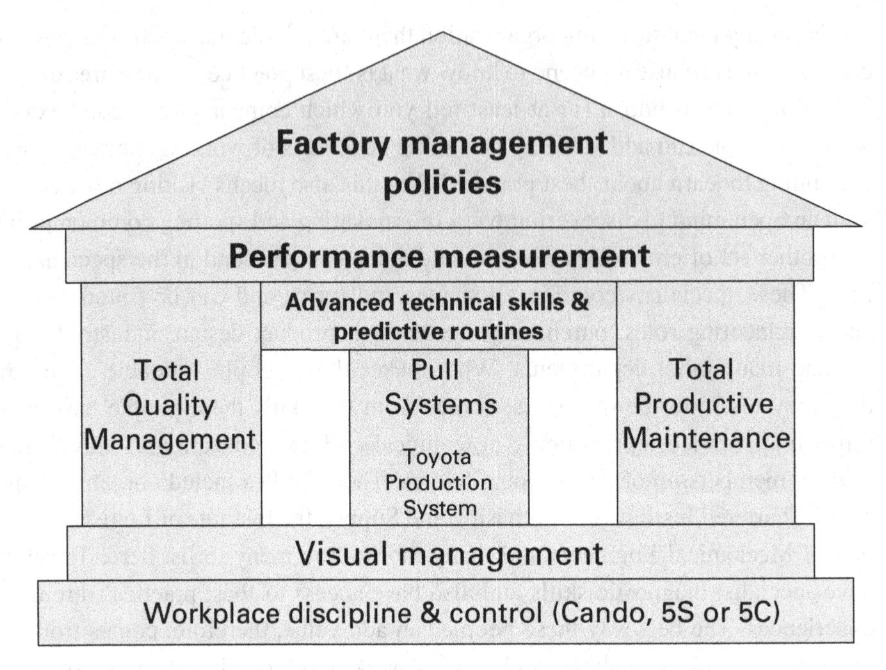

Figure 2.1 The house of lean

that fails to keep in mind the future is not equipping itself for success and these change programmes cannot just be a veneer but a total system of material flow.

The common features of lean business systems are therefore the correct identification of organisational roles and their value in their fight for success.

Having looked at the general organisational roles, we can now move to identify the typical features of a lean production system. It is important to remember at this point that the lean design is there to satisfy customer needs and to achieve perfection – it is not a mixture of all the 'best practices' that could be found and welded together. The purpose is to enhance value – this is the system to achieve it profitably! Simplicity is key and the system must be capable of being understood by every employee and often employees in customer and supplier businesses.

The common features of the system and the operational change process to implement lean production form a natural order that is often presented as a 'house of lean' – much like a Roman or Greek villa. Its an interesting analogy and does suggest that lean production can only be achieved when it is put together correctly and when key organisational processes are brought together to give it strength. The foundations of the 'house of lean' are basic operations disciplines, the floor consists of simple and visual techniques as controls and the walls are produced from quality control, maintenance and material flow pillars, which provide structure and robustness to the system. Finally, to keep everything synchronised and in-place there are the binding measures of the business and the use of factory policy setting to focus and give direction to the many improvement programmes of the factory. It is not enough to have one piece or fragment, the power of a lean system lies in the design of the total 'system'.

So the strength of a lean production system shows up as its total structure and therefore there must be a logic as to where to start and how to put the system together. We will now examine the logic of a lean production change programme.

The logic of lean implementation

This logic is needed because it will be used to gain consensus throughout the entire workforce. That means it should be simple, understandable and that each addition should be seen as a logical extension of what has gone before, not like the 'old days' of 'fashionable techniques' that never worked and were quickly replaced by the next 'big idea'.

The logic of lean implementation is also quite straightforward and common to all companies (Rich, 2001). It should be easy to communicate, should focus on practical issues of relevance to factory workers and should use 'learning by doing' not 'death by computer presentation'.

The stages of a sustainable lean improvement programme all commence with projects to improve the safety and morale of the factory. These programmes (that will be explored later during the chapter on the 5S/CANDO methodology) are highly visual and attempt to integrate workers with the change programme by improving the workplace and conditioning the teams, in an easily understood process, to the stage of problem solving. At this point, the quality of the process will improve, but this is a secondary benefit of a workforce that feels integrated with a visual change process. The visual nature of the change is important, particularly for established businesses, when it is important to show that management means business and this is not just another 'management fashion or fad'. So where better, and least political, to start than in improving the safety and morale at the factory?

Starting at the point of 'adding value' unites the interests of all stakeholders from individuals to the trade union, to managers and customers (who are more frequently visiting their suppliers as too are families of employees). No world-class business exists in an environment with poor safety standards and poor morale – you do not get the best from people if the factory is a mess and the basic organisation is not in place. In companies where the factory is a disorganised mess, it is difficult to tell the difference between good and bad standards of professional conduct. It also often means you cannot see the value from the waste, and you cannot see the dangers that exist in even the simplest of working environments.

A sad reminder of this need to get the foundations right is the sheer number of accidents in factories nationally and the number of employees that now seek compensation.

It's a basic right to expect that the factory in which you work is safe, even though it may not be 'world class' or particularly lean!

The second stage in the logic of lean implementation is to address the 'quality' of everything that is done in the factory – this means understanding the value of doing what you do and how to do it better to provide higher levels of value. This stage, just like the first, is designed to create order and stability from chaos and noise! It is a process of getting things organised and to regain control. Some people will now be wondering how this focus fits with the 'house of lean' – well each of the pillars and visual management practices start with basic problem-solving activities and the engagement of all employees in eliminating waste by being taught how to find it. So naturally, most companies will begin to build the TQM pillar but, in truth, the pull system and maintenance pillars also have stages of problem solving, but tend to include specialists rather than the operations teams. Regaining control of the conversion process therefore means controlling the quality of the process as defects are waste! Defects have no allocated space to be stored in the factory and, therefore, to improve flow, we must eliminate all forms of defects.

Let us spend a bit of time and get this 'quality first' concept into perspective. Traditionally, manufacturers have taken decisions to reduce costs and this has resulted in some completely irrational decisions. These decisions include halving the amount of cement needed to build a solid foundation (or cutting the amount of training for example – a real indicator to employees that managers don't really care about involvement, empowerment and worker ideas!). The traditional approach misses a trick though. The sustainable way of reducing costs is to improve the quality of the production system and the administration of the firm. Improving quality improves productivity and reduces the time needed to convert products and hence it lowers costs. Lean production systems follow this 'quality first' route and unsurprisingly increasing quality improves customer value (internal customers, external customers and consumers). Proof of this can be seen in the success of Japanese products in the automotive and electronics sectors. The key motivation of these businesses was to increase the quality of the product and the processes that made that product and this resulted in market share gains (which is a reflection of customer choice and the 'value' proposition of the manufacturer to the market).

At the third stage of lean logic, you will find improvements in the 'delivery' systems of the firm. This is for a very simple reason, if quality is good and constant, then products should flow better and therefore batch sizes and inventories can be reduced to allow quality products to be delivered in less and less time. For some manufacturers, the introduction of Just-In-Time manufacturing preceded the development of a 'quality consciousness' and guess what – Yes – defects were transited around the factory quicker, which meant they got detected later and at greater cost. This created two problems: the first was that the product was no good and the second was that in making the product vital time was lost and could never be recovered.

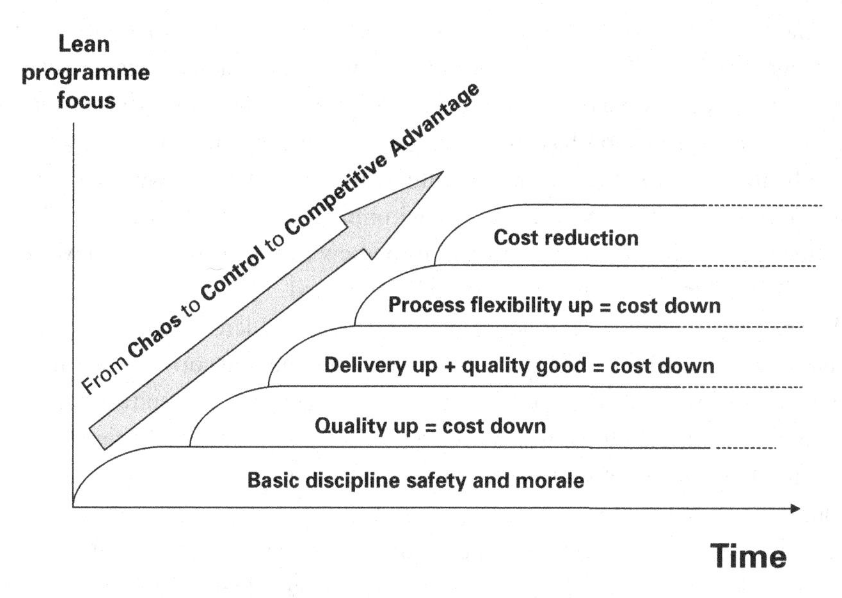

Figure 2.2 Lean improvement stages

Improving the delivery performance at each stage of production therefore shortens the time between receiving and despatching product orders. This is a major 'order winner' in most manufacturing industries and is stated in terms of shorter lead times. In fact it is often the case that customers will buy slightly more expensive products if they have good quality performance and can be delivered quickly. So stage three involves an attention to 'delivery issues' and the introduction of solutions that compress the time between getting an order and getting paid for it.

Stage four is where the issue of costs can be addressed effectively. With high levels of quality and delivery performance from the production system, the only waste that remains in the order-to-delivery cycle is to find ways of reducing any unnecessary costs. These costs can be found in the unnecessarily high levels of safety stock that had been introduced historically to protect against disruption to material supplies. If quality and delivery systems are effective and high performing, then it is safe to reduce safety stocks. At this point wastes and costs become visible as 'abnormal' parts of the production system (lower stock turns and it prompts the use of the problem-solving skills to understand 'why?' these aspects of the process appear unusual). All too often this stage of improvement involves the questioning of policies that surround the manufacturing process and have not been questioned since the system was designed. Therefore cost reductions can take place and further waste eliminated from the production system that was not detected during the previous stages.

In parallel to the cost reduction stage, or shortly after it, is the final element of high performance manufacturing and that is the ability of the production system to introduce and manage a wider range of products. The 'flexibility' stage is therefore used to reset the production system, to compress production batches further and for teams to re-engage problem solving to improve these new products and process standards. The stages and processes must therefore begin again. From the customer perspective, the manufacturing system now has good quality, good delivery and can offer a good product range from which to choose. These features all add value and even requests for the product and factory to become more 'environmentally friendly' can be achieved by refocusing the problem-solving techniques on new forms of 'production waste' as the capability has already been built and can be exploited.

We decided to provide this logic for good reason. Incrementally building a lean manufacturing capability will give local benefits if exploited correctly (i.e. using quick changeover techniques), but, until the basic lean system of quality and delivery performance is mastered, then there will not be a great deal to offer the customer. Just think about it. Halving the set up time for a machine in the production process is actually meaningless until the same savings can be achieved throughout the entire system; until output has increased throughout or boxes of unnecessary inventory removed. It is our experience that, for most companies that distort the logic, there has been little gain from the customer perspective and instead we have made a 'point improvement' in the system of manufacturing, but this has been completely lost as overall stocks have not reduced. It is a sad thing to admit to, but it is nonetheless quite typical of manufacturers, who have experimented with the techniques of lean production, to fail to build the house that supports any given method. The other problem is that, for the bulk of the workforce, the quick changeover programme has little or no meaning as it does not affect them greatly and most will be unaware of it.

On the other side of the equation though, this simple model of 'mastery', to the point whereby the production system is stabilised, is easy to understand. Even at the beginning of the lean programme, it is possible to tell all employees what comes in the next chapter of change and what they will be involved with as the current phase finishes. This has many benefits. It helps to lower fears amongst workers. It helps to convince sceptical workers that 'management has thought it through' and 'it's not a five-minute fashion' and finally it sets the expectations for all levels of employees in the organisational structure. In today's factories it is also quite common to find workers who have experience of the techniques needed at certain stages and therefore employees who can see an opportunity for their own contributions. The latter creates the opportunity for volunteers to be involved in aspects of the production system and to make it their own, and this level of involvement is key to the sustainability of all change programmes.

Chapter conclusions

This chapter has outlined the reasons why businesses must change and improve to fend off the constant onslaught of competition. It has also provided a basic introduction to the lean approach and its origins before presenting a general model of the lean production features. Within the model, it has also been explained that, for the workforce, the process of implementation follows a distinct logic, whereby safety and morale improvements are sought before quality and then delivery improvements. Finally, and at odds with the traditional Western approach, is then the pursuit of cost reductions to eliminate unnecessary business inventories and to reset the problem solving (quality stage) such that new improvements are sought to correct actual and potential failings with the production system. In this manner, lean organisations involve periods of identifying and solving detected failures to maintain a stable conversion process, and also engage in problem solving of issues that could potentially, but have not yet, caused interruptions to the value-adding process. The next chapter will continue to develop the model of lean implementation and the techniques that support a successful and sustainable programme of change.

End note signpost for recommended reading

The first book that should be read when considering the lean approach for your business is the landmark text *The Machine that Changed the World* (Womack *et al.*, 1990). This text first coined the term 'lean production' and is an easy read. The book provides a great introduction to the power and potential of a correctly designed lean system. It is a book that is for everyone – not just 'petrol head' car lovers!

The next text to read is the follow-up *Lean Thinking* (Womack and Jones, 1996) which explains the logic of the lean approach and provides cases of Western non-automotive applications of 'lean'. If these texts are read in this order then a full appreciation of the lean design can be achieved. Other books, which describe the 'logic of lean' for a 'more engineering' audience include Rich (1999) or McCarthy and Rich (2004) and for quality specialists *What is Lean Six Sigma?* (George and Rowlands, 2003). At this stage, no further texts concerning techniques are necessary, as this stage of development is one of 'Awareness and Education' for the reader. The key texts concerning techniques will be explored during the relevant chapters later in this book when these tools can be appreciated in the context and stage at which these techniques can be introduced to maximum effect (and sustainability).

3 Understanding your organisation

Nick Rich, Ann Esain and Nicola Bateman

Introduction

Working for a lean business has many advantages and attractions for every employee. Posing the question, 'what's in it for me?' reveals some interesting answers.

For business directors, the lean business offers a working environment within which there is lots of factual information about the business and its processes together with time, free from 'fire-fighting', to plan the future direction of the business. In the context of the traditional factory, managers are surrounded by 'noise', constant interruptions and chaos. This is not suited to effective thinking and 'direction setting' for the businesses (Dimancescu, 1992).

For middle managers, those who control business departments, lean benefits include working with other managers in groups to improve business processes in a more meaningful manner. This interaction is priceless and improves every managers' understanding of how the business works and of the pressures faced by each department. It is the time spent team working and the decisions of this management group which yield the largest of business improvements and financial savings.

For the operations teams, the lean environment creates stability and discipline within the workplace. It makes work more interesting and there is a lot more emphasis on Improvement, Development, Engagement and Autonomy (IDEA).

What has just been described cannot be achieved by an individual, it cannot be achieved quickly but it must be the goal of all lean businesses. For too many businesses though, what we have just described is somewhat distanced from current factory life. Getting from the 'here and now' to the 'future state' business organisation is an exciting and rewarding process. This process involves an evolution, a process of change which is incremental rather than easily predicted, and it's a process of learning from which most managers get a real 'buzz'.

Unfortunately historic change programmes, especially that of 'de-layering' businesses has changed the way in which the organisation is structured (and roles allocated) without really adding any benefits. Changing the organisational structure, by de-layering, increases confusion concerning responsibilities. De-layering effectively added nothing to business performance. In short, this form of intervention 'thinned out' the management grades. Changing strategies, structures and contracts of employment is therefore a delicate and difficult process. Getting these organisational changes wrong or failing to condition/integrate employees in the change process will, at best, slow change and, at worst, provide a rallying call for all of those employees who want to actively resist change.

A 'Just Do It!' approach ('kamikaze' approach) to improvements will generate confusion for employees as they watch a 'scatter shot' change process that appears to lack logic or structure. During change management and process improvements, clarity of purpose is needed – for it is ultimately the people in any production system which determines its efficiency/effectiveness. To this end, all employees must understand and accept each stage of change of the change management process which builds upon the foundations laid by the last change initiative. Understanding 'why?' we do things is therefore very important as it helps all employees make better decisions in their daily working lives.

Contrasting the traditional and lean organisations

There are many texts that present the 'tools and techniques' of lean production, lean supply, lean design, and much more, but what these texts tend to lack is any real description of the 'enterprise' as a single business system, how it is structured and how it is planned and controlled. The features of the lean system are however very different to that of the traditional Western business and, before we go deeper into the lean systems, it is worth looking at the 'essence' of the lean enterprise model.

The stereotype 'traditional mass production' and 'lean' forms of business seem to be completely at odds with each other and this does cause problems when looking to change from a traditional business model to a lean one. It should be noted that these contrasts represent differences in mass and lean 'ideal types' and in some respects the ideal lean enterprise does not exist in the same way that the pure 'mass production' models do not either. However, senior business managers must understand the differences in 'enterprise design logic', because to understand the 'ideal type' lean enterprise helps inform business decisions and how to organise the firm. To put this into context, a design logic is the underlying view of the organisation, so, for example, if you see a business with large batch sizes, piece rate pay, and the hiring/firing of labour, then you are looking at the manifestation of a design logic called 'mass production' or traditional management. Under this logic, the motivation was to maximise profits, to minimise costs and to stop any form of labour involvement with the business. The old logic was employees could not be trusted and would use any information about the company to increase their pay. Understanding the new 'logic' of lean is therefore very important. Table 3.1 compares the features of the traditional model against that of the lean enterprise.

Table 3.1 *Design logic of the traditional and lean firm*

Business feature	Description	Traditional model	Lean model
Business intention	What is the objective of the firm?	Profit maximisation, cost minimisation	Value maximisation, waste minimisation
Business decision making	Process of decision making in the business	Management prerogative and exclusive management power Command and control	Top–down direction but emphasis on middle management integration Reflexive management Collaboration
Business structure	How is the business organised to meet its objectives	Departments with clear boundaries	Teams in value streams that consume departmental activity
Technical system			
Technology management	Use of technology	Make as much (buy as little as possible) and displace labour	Sensibly automate and make all strategic elements of the product
Production system design	How is material fed through the production system?	Scheduled 'push' of materials through the factory	Product flows with pull systems where appropriate
Batch sizes	What is the unit of production?	Large batches	Single piece flow with quick changeovers
Technology	Specification of assets	Large-scale and dedicated assets	General purpose and flexible machinery
Human resources			
Importance of labour	The view of labour within the firm	Variable cost to be hired and fired as a function of economic circumstances	Full-time core workforce is an asset with 'lifetime' employment Temporary labour is used to protect core workers
Role of labour	Task or process focus	Do as told – TASK Improvement activity is specialist responsibility	PROCESS Empowered workforce and autonomous teams
Contract of employment	Formal agreement with company	Formal and tightly defined roles	Wide, broad, with high levels of flexibility
Remuneration	Employee payment systems	Piece rates and hourly paid	Salary with company-level bonus
View of trade unions	Position of the company towards trade union representation	Arms length. Limit information. Do not trust	Collaborate. Inform. Integrate

Table 3.2 *Traditional and lean supply chain*

Supply chain feature	Description	Traditional Model	Lean Model
Customer integration	Extent of involvement with customer	Avoid integration. Concentrate upon improving margins by 'winning' price negotiations	Understand customer as a means of finding new ways of improving product and margins
	Frequency of interaction	Annual	As regular and often as possible
	View of customers	Short term	Long term
	Range of customers	More is better	Customers selected based upon long-term growth prospects
Supplier integration	Extent of involvement with suppliers	Avoid integration, exercise purchasing power and at 'arms length'	Integrate and share information so that suppliers act properly to keep material and innovation flowing
	Supplier contract length	Short term so that renegotiations are more frequent	Long term but with expectation of annual cost reduction
	Key measure	Piece part price	Quality, delivery, cost and general capabilities of supplier
	Product specifications	Customer drawings and legal penalties for supply failures	Co-design of products and supplier development activities
	Supply base	Vast range of alternative suppliers	Select range of suppliers
	Supplier location	International where the prices are lowest	Local and integrated for Just-In-Time supply

In terms of the overall design and strategies of the firm, the focus of mass producers on profit maximisation, cost minimisation and efficiency have resulted in functional specialisation, whereas the lean business is focused on profitable customer value and the management of business processes rather than departments. The traditional business is dominated by managers dictating and demanding change within a business culture that does not share information nor seek the integration of the individual with the performance of the firm. The 'co-destiny' of employees with the business is an essential element of the lean model and all of its relationships beyond the factory. Here too there are many differences between the traditional and lean businesses (table 3.2).

The emphasis on an 'integrated system' of management and supply chain processes therefore unites the activities of the lean business. For traditional firms, the dislocation of customers, suppliers and the workforce creates a fragmented system within which there is an information vacuum. In this vacuum there are many opportunities for the traditional business to be sub-optimised. Departmental silos act independently, often ignorant of the problems that decisions in one department have upon another, or indeed the key customer service processes. The lean system is less introspective, less fixated with departmental thinking and is focused upon growth and harmony within and beyond the business.

For most managers, it is relatively easy to select the model of business that they would wish to pursue and adopt, but again, for most managers, there is already an existing system and structure which prevents making this choice simple. Changes to the structure of the business, the application of lean tools and techniques is comparatively easy, but to achieve these changes, whilst managing to adapt the 'essence', structure and culture of the business, is much more difficult. These broad issues of structures and 'indirect' processes are important and something that the entire management team must control and reflect upon. The ability to change these features is not easy, it can be subject to opportunism (such as the changing of formal contracts of employment as prompted 'working hours' regulations), and it involves a series of evolutions (in the right direction) to finally configure a system within which all features mutually reinforce to create an integrated business system.

Company managers in this evolutionary process between these two business models will know well the issue of 'perceived conflict' between 'what is' and 'where the company wants to be' in terms of its structures and processes. These perceived conflicts are readily identified by people within the business and deliberately exploited by some employees who are intent on causing trouble. These will include issues of being paid for new skills that change the current contract of employment of the individual, whether payments can be expected for being trained, whether 'last in first out' layoff policies will be maintained, and even the conflict between local sourcing of product requirements and buying in the same materials from low labour cost regions. There are more such 'hot spots' as the strategy, structure and processes of the organisation 'transitions' the business towards the lean model. It is for this reason that senior managers must lead the company-wide change programme and drive the process in a top–down manner, whilst being sympathetic to the time needed for individuals to get used to the requirements of the 'new job'.

The change process

From our studies we have learned much about the process of change – a lot of our learning and experiences do not reflect the neat approaches taught during MBA classes either. Our summary of lessons is shown in table 3.3.

Table 3.3 *Lessons from lean programme successes*

Lesson 1: Set direction
1a: Understand your position in the world.
1b: Recognising and articulating the 'need and the gap'.
1c: Setting the challenge.

Lesson 2: A business is a system
2a: Recognise the organisation is a set of dependency relationships.
2b: If measures drive the behaviour of managers, then measures affect the willingness of managers to act as a team.
2c: Suppliers reflect the way in which they are managed.
2d: Managers have the power to design systems and know where to find 'best practice'.

Lesson 3: There is no such thing as a bottom–up strategy
3a: People enjoy working in teams – teams add value.
3b: Learning by doing follows the design of learning.
3c: Past business culture affects how you get things done and the pace of change.

Lesson 4: Lead with confidence

Lesson 1: Set direction

The first lesson we can offer is to set and remain committed to the 'direction of change' needed from the firm and to closing the competitive gaps the firm has between the competition. This process includes understanding the current performance of the firm, setting out the challenges that the business will face in the future in a manner that is meaningful to all employees, and to allow employees to select the most appropriate programmes of improvement within their part of the business. The latter can be the hardest part of the lean transition process for managers, who are used to telling others what they want rather than trying to enlist people and to unlock their creative potential. This means asking others to study the business and its current situation, to find out what is going wrong and to work with others to propose key company-wide initiatives. These initiatives are likely to take many months to implement and will consume a fair amount of business resources – these are the breakthrough projects needed to make the business competitive. This 'commitment', by senior managers takes confidence and marks the distinction between the manager as a 'distant figure' and the manager as 'a coach' to the people they lead.

Lesson 1a: Understand your position in the world

The first stage in the process of change is therefore to understand the 'world around you', and it is not a difficult process. For most businesses, it is the customer that has dictated the need for change and a general excess of supply over demand. These pressures have lifted the competitive stakes. Customer expectations of service, quality, costs and the general rate of product and service innovations has lifted. This market driven change has therefore reduced lead times from many weeks to days and now to hours. It has seen quality measured in percentages, then parts per million and now in Six Sigma levels of quality. And finally it has seen producer pricing, where the manufacturer added a margin to costs and then offered the product to the market at that price, move to a modern focus where the customer expectation dictates the price that is 'acceptable' (market pricing). In this context it is harder to differentiate the product/service offering by the firm from its competition.

Even monopoly industries have been exposed to new forms of competition and have not escaped the need to improve or face decline. In these sectors, state businesses have been privatised, deregulated and exposed to market pressures and even those that remain as monopolies are now heavily audited to ensure that 'profiteering' is not used to the disadvantage of the consumer.

Understanding the world around you is therefore quite a sobering exercise, made more uncomfortable when you take a dispassionate look at your business, i.e. as if you were a customer. You can now understand where the gaps between the performance of your business and customer expectations vary. At this point it is also a little embarrassing to compare the business strategy and current key improvement programmes with customer expectations. More pointedly, many managers will be concerned that the business does not have a strategy, what is in-place is not effective or the current strategy lacks 'customer focus'. This is quite a common experience and few businesses have a robust strategy that has been written down to help every employee to understand the challenges of the market and how the company intends to be the 'best it can'. It should be remembered that, whilst 'stakeholders' are important, it is selling products to customers profitably that generates the dividends for shareholders. All too often we forget this in the UK and spend more time worrying about shareholders at the expense of customers.

Lesson 1b: Recognising and articulating 'the gap and the need'

Managing the evolution from existing business model to the lean design must be based upon a 'need' and a compelling reason – the most compelling is that the business is in decline. This is known as the 'burning platform' and it is a means of convincing everyone to change, but it has its limitations. If every year there is a burning platform, then employees will stop believing their managers. A consistent message which promotes the need to improve is one thing, but, when the top management of the business continue to portray a business in decline, then key people will leave (if only to protect the security of their families).

For most Western businesses, the key 'need' is to reduce operational costs to remain competitive and to achieve this by improving the quality, delivery, and 'time to market' processes of the firm. These improvement-focused priorities have remained the common themes promoted in Toyota since it won the Deming Prize in the early 1960s. Making more of what the customer wants with less resources in a Just-In-Time (JIT) manner is the essence of the lean production system developed at Toyota. For Toyota, a fledgling business in the automotive world of the 1960s, this focus on improvement and the elimination of waste has served the business well – which in early 2004 became the second biggest car producer in the world.

Articulating the 'need' for change and consequently what the 'end game' for the firm (what the business looks like when it achieves its strategies) represent key elements of the senior management role. After all, why should anyone else in the business, especially the majority of operational staff, 'buy into' your analyses of the future and understand the gaps which need to be closed if customers are to be retained and attracted to the firm. To cut things short – the need is to expand the business because growth brings security for all employees.

Lesson 1C: Setting the challenge

It is important to determine the current performance of the business and also the changes in these performance measures over the coming three-year period. These measures should be related to customer service and importantly the processes of quality, delivery, and cost management. Conducting such an analysis is important both as a means of predicting the rate of market change and to set these three-year targets as stretch goals or challanges for the factory improvement teams. For one aerospace engine manufacturer, the key improvement challenge was to '*reduce to 45 days its engine build process*'. For other businesses it could be to improve quality levels by 50% or to find ways of reducing the current product costs by 10% in order to meet an expected reduction in sales price.

These challenges can be understood by all personnel and setting a challenge to the entire workforce is an interesting departure from the traditional process of change within which the boss dictated what to change and then allocated the task of making the change (not finding out whether this was the best change to make) to a subordinate who then in turn allocated to others. The command and control system resulted in written personal 'objectives' to implement what the boss told you to do – a process that neither requires nor taps into the creativity of people in the organisation, it just demands you do what you are told to do. Now imagine if you asked everyone to find ways of reducing the lead time of your product – imagine the amount of different ideas you would get.

As we have already said, a strategy is about satisfying customer profitably by using your human resources in the best manner – these human resources are the very people who compete with the people in your competitor's business. A challenge unites them all and helps reinforce the 'people focus' that is nurtured by all 'world class' performing businesses.

Lesson 2: A business is a system – supply chains are just bigger systems

Whilst questioning and disagreements are signs of a healthy organisation, destructive political conflict is not. Yet many of the traditional approaches to management stimulate conflict, much of which results from poor understanding between employees. There is the conflict with customers, where traditional theories proposed that manufacturers should profit maximise and use every tactic possible to increase prices. There is conflict between managers seeking to gain a share of an increasingly smaller overall budget. There is conflict between managers and trade unions and there is the conflict to 'screw down' supplier prices. It seems that the traditional business model – focusing on cost minimisation before market pressures later replaced this motive with the measure of 'customer service' – was riddled with conflict. The modern approach is to collaborate in order to optimise the performance of the system, be it the supply chain which services the consumer or the internal flow of work through the factory. Conflict is the inevitable outcome of 'fighting' for your department in order to achieve improvements in the costs of your activities (without worrying too much about your actions on others). In this manner, quality managers were often 'called to account' for poor product quality, which, when investigated to find the 'root cause', resulted in decisions by the purchasing department to lower piece part prices by switching to cheaper but less dependable suppliers.

Lesson 2a: Recognise the organisation is a set of dependency relationships

All managers are dependent upon all others is a major learning point when setting direction towards lean production and the lean enterprise. If, in our example, the quality manager's performance is dependent upon the purchasing manager, then the operations manager is dependent upon the production control manager, who in turn is dependent upon the engineering manager and the data collected about the process. When it comes to managing the key processes of quality, delivery and the costs of existing and new products you find many dependencies which must be acknowledged and exploited to yield business effectiveness. Any manager who sees no dependency upon the performance of other managers is delusional – even accountants need data about the process, materials and capital equipment in order to do their job effectively. Following an order as it passes through the different departments of the factory should be enough to convince managers that dependencies exist and that slippages between departments is a major form of waste (which the customer sees and pays for).

Lesson 2b: If measures drive the behaviour of managers then measures affect the willingness of managers to act as a team

Often the measurement systems and key performance indicators (KPIs) of managers have evolved historically and have not been checked to see whether they do reinforce – as they are meant to – the right behaviour within the business. Measures tell managers how well they are doing against control items that measure key processes. In the past these processes were 'departmental' activities, whereas today processes are 'end-to-end' within the business, and departmental measures often create conflicts between managers unnecessarily, especially where these measures emphasise the cost rather than the quality and the delivery of the process. Imagine a business within which the purchasing manager is measured on piece part cost and the operations manager is measured on machine and labour utilisation. Guess what, without constraints these managers will buy in bigger and bigger amounts and then they will process it all in bigger and bigger batches. Queues will grow, whilst work in process rises, customer service falls and the inventory, quality, production control and finance managers will all begin to complain that the business is going out of control by their measures. In a world where all managers must pull together to make the entire system effective in profitably giving customers what they want, revisiting the measures of the business and its departments, to make sure they still represent good indicators of performance, is a worthy activity. Measures show us how well we are doing and should be devised to control key elements of business performance – for lean enterprises these measures will be process focused and shared with many other business managers along the value stream.

Lesson 2c: Suppliers reflect the way in which they are managed

There has been much talk about Toyota's supplier policies and the approach to collaborative buying. These policies have been portrayed as buying everything from one source and spending a lot of time developing suppliers. This is only in part true. Toyota rarely single sources all of its production to just one supplier but instead ensures that there is one source for a specific part and that an equivalent part for another vehicle is provided by another supplier. Without a guaranteed supply of materials, the lean system quickly stops and results in zero customer sales. The issues for Western businesses include how to rationalise the supply base to lower the complexity of supply, how to integrate suppliers often after a history of adversarial purchasing and how to tap into the creativity of people in these businesses. Whilst these issues will occupy the minds of purchasing and supply chain managers, what is happening is there is a growing awareness that suppliers are extensions of the factory, producing parts that the business cannot or does not want to make. As extensions of the factory it is not sensible to remain in conflict with the supply base, to withhold information and not to develop suppliers. In effect they are external departments, which also share the common interest in the quality, delivery and cost of the materials/services they provide. Whether we care to recognise it or not, every business is dependent upon the performance of its suppliers – 85% of a motor vehicle is made by these suppliers. Furthermore, constantly 'beating up' suppliers has proven a poor means of getting suppliers to change and become more 'customer focused' themselves. In parallel, if a business is dependent upon its suppliers, then the same holds true of customers and it has never been written that a supplier cannot and should not develop their customers (or at the very least understand what they 'value').

So we are dependent upon managers within our own business, we are dependent upon managers in supplier businesses and also we are dependent upon managers in customer businesses, and, if we are to optimise the business and its supply chain, then these relationships must be recognised and nurtured. Collaboration is therefore preferable to confrontation if the necessary relationships are to be developed effectively and supplier businesses are to build faith and trust in the customer business. If all suppliers are aligned and working with the customers to find ways of making the product in a less expensive or more innovative way then the system is working for everyone. Suppliers get more business by helping customers to develop profitable and attractive products.

Lesson 2d: Managers have the power to design systems and know where to find 'best practice'

Many managers have a 'wish list' of what *they* would change in the factory but no real understanding of how this would affect other managers – even though they are, as we know, dependent upon these other managers. In terms of the design of the organisation, departmental managers have the organisational power and responsibility to design their own systems. The boundaries of departmental activities are only the result of where the lines of the organisational chart are drawn. Managers can therefore work together if they choose to – it is neither a sign of weakness nor indeed a sign of being subservient to the other manager. The simple truth is that, without collaboration, most major businesses' 'step change' programmes cannot and will not happen. And that is a real shame because departmental managers are professionals and these professionals all promote 'best practices'. 'Best practices' including Six Sigma, lean, value engineering and many others all require cross-functional management collaboration to implement and sustain them. The traditional organisation is therefore limited in its ability to introduce changes of this nature because of the lack of management collaboration and cohesion. Even business departments that have high levels of political power cannot 'force' change that sticks.

The key learning point, from our experiences, has been to allow middle managers to get together as a group to discuss 'best practices' but in the context of discussing business weaknesses, such as excessive lead times, quality levels worse than competitors, or higher costs. In this manner, managers can promote their knowledge as a means of helping the business and of helping other business managers achieve their goals.

Lesson 3: There's no such thing as a bottom–up strategy!

The importance of a good middle management group (departmental managers) who have mutual respect and feel comfortable working together (they trust each other) is a characteristic of 'world class' businesses. These employees tend not to naturally bond as a group, and therefore it is important that group sessions are used, at which attendance is compulsory, to study the organisation, to engage in mapping activities, to visit other businesses and also to learn how to meet the challenges of the firm. At these sessions, the dependency of managers upon each other must be stressed and simulations (team games) used to integrate the group – no matter how personally uncomfortable this is for certain managers at the beginning of the process (these feelings soon disappear with practice).

Middle managers are the driving force for change, especially when presented with a challenge by the senior management team to transform the business and are then left to do just this without excessive senior management interference. These levels of the firm set the direction and pace of change despite the promotion, in contemporary management texts, of a 'bottom–up' approach to strategy. This is a mistake – shop floor teams know the most about the tasks they perform but can be terrible strategists, and, if they are good business strategists, then why keep the senior managers? Shop floor teams should keep control of processes and report accurate information so that the senior managers know at what levels they are performing as a business and can, therefore, examine the gap between present performance and where the business needs to be to compete with the best in the industry or the toughest customer expectations. But strategy formulation is not for the shop floor – participation and enactment is. The latter process is inextricably linked to the process of policy deployment and the use of stretch goals to focus improvement activities within the firm.

Lesson 3a: Teams and adding value

Managers, shop floor employees, and indirect workers, add the most value, through improvement activities, when they are part of effective teams. Its human nature to become a team member, but it is an organisational management process which is ignored or treated lightly by traditional business managers.

For the lean business, teams are the basic building block of the firm – an element of the lean business system that is taken seriously. Trade mark features of team-based high-performance businesses include: problem-solving groups, cellular working, cross-functional management teams and supplier clubs. These teams, permanent or temporary, lower risks and stimulate innovation if the teams are armed with the right diagnostic tools and are experienced in their usage. However, the formation and management of teams is not easy and groups that are artificially put together by managers, with little thought to the social dynamics and technical skills of the group, will add stress to certain individuals. Planning and not putting too much stress on embryonic teams is important – so managers must be present to help team development.

Lesson 3b: Learning by doing follows the design of learning

For all of the benefits of learning by experimentation, most managers and line workers have long memories when it comes to changes that have failed – especially line workers. Everyone knows how difficult it is to learn a new language and just getting your mind around what is lean production is difficult without many of the Japanese words that describe particular practices. It's confusing at the best of times. However, it needn't be. The art of good lean implementation is making the learning, doing and embedding process of change simple and logical.

Unfortunately we do not always value the simple in the West, in favour of believing in the one complex breakthrough solution. But in truth, lean is more about the quantity of simple solutions ('small steps') rather than 'big bangs'. The complex are not easily understood and therefore, if the people tasked with making the process work cannot understand the 'breakthrough', then it is unlikely to be sustained. Making change practical, less intimidating and 'user friendly' is an art. The second art of lean implementation is to work out a set sequence of improvement activities that provide logic to the operating teams. By 'logic' we mean a sequence of events and learning experiences, where the next lesson is taught when the last lesson has been sustained and the next lesson represents a natural and very logical extension of the last. Jumping around from tool to tool with no reason is frustrating for operating teams and it does not reinforce why the technique has been selected. However, if you start simple with improving the workplace and then move into problem solving and finally allow the problem solving to tell you which improvement techniques to select, then 'it all fits' and individual team members can see the logic that took them from problem identification to the countermeasures and solutions. Such an approach to change is 'local' to the team within the factory and helps the 'ownership' of the change process itself. Making things simple and logical is a real art and a profitable one when it comes to sustaining change.

Lesson 3c: Past business culture affects 'how you get things done' and the pace of change

It would be ludicrous to believe that a few well-chosen words and promoting the principles of the lean enterprise and lean production systems within the factory will be greeted with resounding applause by managers and other employees. The willingness of a business to change is based upon their perception that the business is willing and capable of helping the employees. These perceptions are based upon the historic experiences of change that the workforce has witnessed.

For businesses that are capital intense (employ technicians rather than operators) or have a low regard for the capabilities and responsibility of operational workers, then there may be a different approach to lean than the typical one. We have witnessed businesses that have spent a great amount of time on the 'human side' of the business, training and involving all employees in some form of change, but we have also seen companies with cultures so 'hard' and anti-team working that have lean systems. These businesses tend to use elites of engineers to design and implement robust systems and then the new system is policed and enforced such that penalties are instigated against those employees who breach the new ways of working. Both the classic lean and the enforced lean systems work – the latter tend to suffer issues of poor morale and low levels of company-wide innovation. It may be the case that current 'elite' systems will evolve into more classical types.

Box 3.1 Buying into lean

Buying into lean – one organisation's approach to change

A phone call to LERC resulted in an interesting conversation with a globally renowned brand. Their question, *'Could you recommend someone who we could employ on contract for six months to become lean?'*

Unfortunately, this is not an uncommon approach to change! The typical characteristics of this type of change are that it has to be quick even though it has taken years to develop elsewhere. Also, the approach involves using someone external to the organisation so that if anything goes wrong they know they have a scapegoat, and management can abdicate the responsibility of leading change.

Learning points

- *The person at the top of the organisation has to be seen to be actively supporting the programme.*
- *The organisation has to own the change and this can be done by appointing a key member of the organisation to champion the programme and gradually engage more personnel as the programme progresses.*
- *Demand fast improvement, whilst seeing it as a long-term approach for the organisation as it becomes part of the new way of working.*

Lesson 4: Lead with confidence

Having a strong leader is one of the clear messages from the book *Lean Thinking* (Womack and Jones, 1996). A leader who supports the change process and 'walks the talk' is necessary to give credence and 'room for doing' for employees throughout the business. So often executives like the idea of lean but do not want to change themselves. It is also true that senior managers often think they need to be lean experts – they don't but they do have to agree the direction of change is correct. However, many senior managers will be impatient for change, and to some extent that is the nature of senior managers, but a lean transition requires leaders who understand that change must stick and that it is really easy to confuse the operational employees if change is hurried. Weak leadership of the lean transition and poor accountability to the most senior managers of the business is however a lesson in disaster. All too often we have seen senior managers who have not bought in to, or do not take responsibility for, the change process – if senior managers cannot see the point, the teams they control will not get it either.

Box 3.2 Establishing a clear message

Keeping a clear and consistent message on improvement

- *Improvement initiatives require good leadership, including managing exployees' expectations.*
- *Support the teams carrying out the change.*
- *Publicise the successes of people and projects.*
- *Clearly communicate what and when an area is the focus for improvement – be open and honest.*
- *Initiate visible shopfloor display boards that indicate the targets and improvements with key business metrics that have been agreed by those who are being measured.*

Confidence is a state of mind which accepts that the precise nature, stages and outcomes of change cannot be accurately determined, but that the direction of change is right for the business and its customers.

It should also be noted that the broad nature of the lean enterprise and lean production systems means a collective approach to senior management leadership is needed. Few managers have the 'power' to authorise and begin such a process of radical change on a company-wide scale. It is not that surprising, given the pressures of the 'day job' and the size of the 'transition problem' in hand, that just 'beginning' the process of change is daunting. Some businesses will even fail to get going or resort to a huge spend on consultants – abdication of leadership and passing the responsibility of enterprise design to outsiders is not a cheap alternative for management. 'Walking away' from the task of leadership is frustrating and managers of this type miss out on a lot of interesting activities, mapping sessions and experimentation which is an essential experience for managers to really 'get into' lean.

Implementing lean systems

The key to the implementation and transformation to a lean production system is the consistency of objectives to continually improve 'quality, cost and delivery' of product to the customer. These issues unite the two aspects of effective change management: the management team (designers of new processes) and the operations-level teams (improvers of new processes).

Implementing lean systems: the management dimension

This book does not attempt to provide all the tools needed for the journey to a lean transformation and it does not defend the view that there is a 'one way that fits all'. It's not true – every lean implementation is, for all intents and purposes, unique. That does not help managers seeking to answer the most basic question of 'where to start' though. This was the motivation and research challenge which resulted in the provision of various value stream mapping techniques to help managers at the beginning of the lean transition.

The purpose of these maps is to create a 'wake-up call', to get employees to question current practice, to identify what needs to change and also to indicate which tools and techniques are likely to be required (and hence what training is needed). It is a shame that most managers pass the responsibility for mapping to consultants rather than doing it themselves, as the process is enjoyable, rewarding and an enlightening experience. Much has been written about mapping and it is an essential part of lean with many different tools to help diagnose your production system from the design of the supply chain, to process design and all the way to task design. Figure 3.1 summarises the relationship between the hierarchy in the organisation and who should be doing what and which tools to use in terms of understanding your organisation.

From our simple model you can see that executive team members must articulate, demonstrate and reinforce the need for change over the coming three years or so and demonstrate a role as leaders to unblock the path ahead. Middle managers must set aside historical conflicts and think about how best to lead and sell change over the next year and the front line supervision must keep the value adding process working, whilst also looking for new areas of improvement that will keep the performance trends of the area heading in an upward direction over time.

So maps help to define the value streams of the firm and the activities that support the primary principle of 'understanding customer value'. Without the mapping process, managers tend to focus on the 'short term' and 'today's chaos' rather than thinking through why the system is failing and what can be done about it. The value stream

What		How
Leadership **Reason & need** *Why should this organisation undertake change – What is the problem? What are realistic expectations? What measurements need to change?*	Executive Unlearn to Relearn Forward 3 years	**Leadership** **Belief & demonstration** *Consistency of message & walking the talk* *Expectation depends on resource availability, pace people can change, product cycle time, etc*
Leadership **Embeddedness** *Provide framework of logic for how to resolve the problem (s)*	Management Unlearn to Relearn Knowing to Doing Forward 1 month to 1 year	**Leadership** **Logic** *Resource – to demonstrate commitment* *Co-ordinate and control – for ownership & measurement*
Control participation *To agree problem and assess different solutions to resolve the problem* *Implement, learn and adjust*	Supervisors and Operators Unlearn to Relearn Knowing to Doing Daily – Weekly – Monthly	**Measure and adjust** *By creating feedback loops against action/ implementation*
Where are we now?	Value stream mapping	Where do we want to be?

Figure 3.1 Who's doing what?

focus and the knowledge gained through mapping generates the facts required to make informed decisions about how to improve the business.

The evaluation of an organisation's value streams is normally undertaken using a mixture of mapping techniques depending on their appropriateness. To be consistent with lean thinking and Ohno's seven wastes (1988) we can use a process called the 'Seven Value Stream Mapping tools' (Hines and Rich, 1997) illustrated in table 3.1 to help us root out waste. The table indicates the applicability of each of the mapping techniques for identifying and evaluating each type of waste. The tools are also described in the booklet *Going Lean* (Hines and Taylor, 2000) from an implementation perspective.

Initially we would recommend the use of at least the maps to understand the make up of the process being mapped – the triangulation of the results will help in the analysis and outcome.

To help the reader, the maps can be defined as follows:

1 *Process activity mapping* is a traditional industrial engineering tool which follows a vital piece of raw material through the production system to the finished goods store. Each stage of the passage from raw to finished product is classified as an operation (value adding conversion) or a waste (transport, inspection, delay, and storage). The map is used to find the value added as a percentage of the total time the material takes to work itself through the production system. The average outcome is about 5 per cent of the time the material is being converted.

2 *Supply chain responsiveness matrix*. This map plots the amount of inventory at every stage of the production process (cumulative inventory in days) against the time (in days) to plan and move the materials. The result is a graph of where the slow-moving stock resides and where to look to make improvements in material flow.
3 *Product variety* funnel follows a standard raw material from the beginning of the process to the finished goods to establish how many different products are made from that material and to establish the point at which the variety begins to expand rapidly. This is the point at which it would be possible to maintain a generic stock buffer and to then finish products to customer order. As such the map has many affinities with the decision point analysis map.
4 *Quality filter mapping* simply shows, for each consecutive stage in the production process, how much waste is generated in terms of scrapped product, rework or service defects. This knowledge allows managers to focus on where to make quality improvements to eliminate waste.
5 *Forrester effect mapping* shows all the problems and delays with schedule information and the production of materials to meet this demand. It is shown as a line graph with a line showing the customer forecast for each week, the actual shipments to customers, the launch of production batches and when the key raw materials were scheduled. These various lines often show big distortions as customers inaccurately schedule, batch sizes are excessive and minimum-order quantities are high. The perfect lean system would have a series of very flat lines showing the order, the making and the re-supply of products (see table 3.4).
6 *Decision point analysis* is another great map which shows the difference between the production times through the factory and the waiting time that customers will accept to identify a point at which the business could 'make to order'. So, if customers will wait only two days as a lead time for products and the product itself is made in two stages, each taking two days, then the map would suggest it could be possible to finish the products through stage two and that stage one and two can be decoupled using an inventory buffer. It is a good tool when looking at what could be done to reduce lead times for customers.
7 *Overall structure maps* are interesting in that they show the numbers of companies engaged in the distribution channels of the firm (and the value they add) as well as the number of suppliers at each stage of the supply chain (and the value they add). These are difficult maps to compile, but quite enlightening.

Whilst none of the tools provides a robust answer to management questions, the trick is in the combination of maps which increasingly narrow down the key issues that need to be addressed. So, let's say the process activity map shows that the factory has low levels of value added and high levels of storage, this is confirmed by the supply chain map and then the Forrester effect map shows that the source problem concerns poor customer schedules and the decoupling point suggests it is possible to lower the lead time for customers (and hence the need for forecasting). It is therefore

Table 3.4 *Waste and appropriate maps*

Mapping tool	1	2	3	4	5	6	7
Wastes/Structure	Process activity mapping	Supply chain response matrix	Production variety funnel	Quality filter mapping	Forrester effect mapping	Decision point analysis	Physical structure (a) volume (b) value
1 Overproduction	H	H	L		M	M	
2 Waiting	H						L
3 Transportation	H		M	L		L	
4 Inappropriate processing	H		M	L		L	
5 Unnecessary inventory	M	H	M		H	M	L
6 Unnecessary motion	H	L					
7 Defects	L			H			

Note: Key H = High correlation and usefulness, M = medium correlation and usefulness, L = Low correlation and usefulness
Table 1: The seven value stream mapping tools
Source: Hines and Rich, 1997.

possible to find key improvements worth lots of money through this form of mapping diagnosis.

Most recently, and an extremely popular mapping approach to emerge, is that of '*Learning to see*' value stream mapping, figure 3.2 (within the company and from 'door to door') and this addition to the mapping family is very powerful. The approach is very graphical and shows a complete manufacturing system from customer (top right of the diagram) to supplier (top left of diagram) combined with the physical process (bottom of diagram) and the planning system of the firm. Initially managers will map the entire process to present 'the current state' of the product family or highest volume product. The next stage is to add to the map all the issues which prevent improvement and to focus on these as key change programmes in the factory.

The other activity, involving these maps is to create a 'future state' and ideal production system. This activity begins with an understanding of the customer demand rate for the most important product family to the business (this demand rate is known as the 'takt time'). The next issue concerns whether the company can flow materials to the shipping department or whether a finished goods stock is needed, then working backwards along the value stream, to identify the points where supermarkets will be needed and also which process is the 'pacemaker' (the one which is scheduled).

To create these maps and to focus improvement activity, a cross-functional group consisting of representatives from the key functions are brought together to visualise their processes. These maps should not need sophisticated computers to create, but

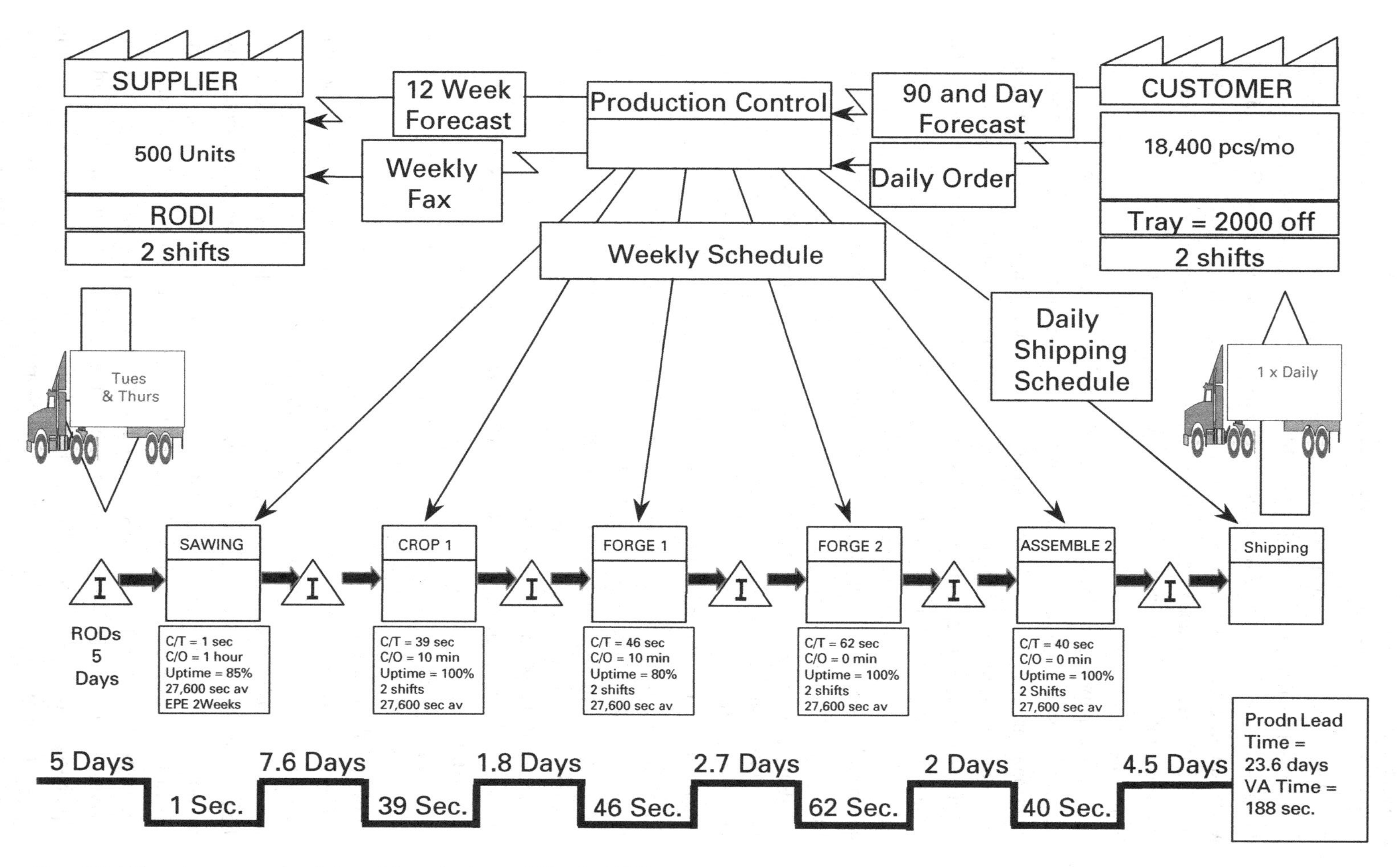

Figure 3.2 Value stream mapping (adapted from Rother and Shook, 1998)

should be conducted using flipcharts or 'brown paper models'. The most common problem when mapping is finding reliable data in a format that can be used, and, therefore, the team are urged to collect such information themselves. These maps make waste identification easy and offer a series of simple improvement activities that will improve the performance of the firm without consuming precious resources (the 'low hanging fruit' of kaizen activities at points along the supply chain – also known as 'point kaizen').

Once the mapping has been carried out, the decisions have to be made as to the key areas for process improvement that will deliver most to the organisation. The cross-functional groups that have participated in the mapping have learnt the purpose of each map and its application and can make an informed decision on their selection and develop a greater understanding based on their findings. Consensus is then required for the action plan, which focuses the improvement activities and establishes responsibility for key projects. The improvements activities should reflect the strategy for the organisation and its key objectives, such as improving customer quality levels, reducing lead times or cutting out costs.

Knowing what to do to improve is different from change and action. Also what is obvious to managers and those who have undergone the mapping process is not necessarily obvious to those who have not taken part. So it is at this point of lean transformation that the management of the business must devise a structure for business improvements and begin to invest in this business capability. The first process that must be designed is how the senior managers can lead change through becoming 'change champions'. This role is to be the figurehead of a change programme, not to get involved in doing necessarily, but in facilitating the process and ensuring that hard questions are asked of the team. The position of power of the senior manager means that they can 'unblock' problems and promote change throughout the business. Whilst being a champion is on a 'project-by-project' basis, there are always likely to be on-going projects and therefore being a champion will become a permanent feature of the senior manager's role, albeit taking only a few hours per week to conduct. This is not an unreasonable request.

The next organisational role is that of 'change agent' and the person that will facilitate the on-going work of the team as they pursue their projects. These individuals are lower in the organisational structure and spend the majority of their time on projects. Both the champion and the change agent (operational project manager) must continuously promote the lean message and in our Learn 2 network the change champion/agent tends to have a weekly or monthly tour of the workplace, allowing the workforce to communicate their lean initiatives and to illustrate their achievements, including their expected returns to the business. More often this is conducted using white boards and flip charts in the workplace, rather than the 'entertainment' of cartoons and moving images in computer presentations, where the improvement is lost as people concentrate on the animation. Problems are also discussed during the tours, which open up the opportunity for constructive criticism and potential ways of finding a resolution.

These sessions have also been used to acknowledge team efforts and reward improvements, so as to reinforce the positive commitment for change on behalf of the company. The final job role – often a role that has a blurred boundary with the change agent – is the establishment of the Lean Promotion Office (LPO). For large and for small companies, this is a distinct office which is populated by employees who have received training in the tools and techniques and who are capable of training others. The LPO also acts as a library for reference by employees. The office holds the standard project charts, documentation and records team successes. At this point, the managers have now understood where lean could take the enterprise, the problems and opportunities that the maps reveal and have a structure to support the initial deployment of improvement teams (Rother and Shook, 1998). The next stage is to understand the change process from the team perspective.

Implementing lean systems: the operating team dimension

Rarely will you find a company that has achieved long-term competitive advantage or world-class performance on its assets or its products alone. It is people that can innovatively and creatively generate solutions, and people, as lean principles indicate, are the businesses most important asset. However, people are the most volatile and sensitive of all business assets and people learn at different rates and in different ways. As such, the design of the implementation and training processes associated with lean production are critical to ensure that individuals learn the quality of the tools, techniques and solutions they use, before allowing individuals to increase the quantity of solutions they identify.

The process of team-based lean production implementation must have a logic that can be readily understood by all in the factory. Further, each additional improvement activity must seem to build on the last ('a logic'). That means managers may well have to engage in activities that add no commercial benefit in the short term, but do impact positively upon employees and condition them for greater levels of involvement, participation and self-management. In short, managers must engage in promotional activities in their leadership role and also stage the improvement process such that good-quality solutions emerge, not the typical approach which is to demand a quantity of improvements ('the scattershot' approach). From our research, we have found that there are three stages that work well as a change model.

Box 3.3 The change model

Change requires three stages:

- Awareness raising and education
- Implementation
- Sustaining changes

The model relies upon the management of the firm identifying the importance of change and providing education such that employees can understand what change means and the tools needed to achieve it. Armed with a general knowledge of lean, most employees will make reasonable decisions, but training is a means to an end; that is, it is only of value when the training is for a purpose and can be implemented. In some organisations training has become part of their 'investors in people' approach to improve the skills of their workforce, an admirable approach but one that is too quickly lost if there are not the opportunities to utilise new capabilities through practice.

As such, teams should not be bombarded with training sessions, but instead the teams should receive training when there is a need for the improvement technique. In Japan, this approach is often associated with the martial arts expression that 'the teacher will appear when the student is ready'. It is an important learning point and suggests that only when teams have a competence in an improvement technique should they be allowed to progress. The latter concerns the process of sustainability to which we will return.

The process of team-based change is the subject of the next few chapters of this book, but, in short, for the teams all change starts with workplace organisation (CANDO) and progresses into higher and more technical levels of problem solving (Rich, 1999). During the problem-solving stage, the LPO will be expected to engage in training that is relevant to the team and the specific problems they face (such as long set up times of machinery etc.).

With these mechanisms in place, the operations teams have the ability to improve their own performance and the senior managers have an understanding of how the 'production system' in its entirety (from 'door to door') can be improved. But what is missing is a means of focusing the improvement process as an on-going part of business planning and change management. For lean companies this is the role of 'policy deployment'.

Implementing lean systems: joining the system together

Legions of managers have made industrial pilgrimages to Japan. These managers have been intrigued by the physical, tangible and simple nature of techniques such as 'kanban' production, mistake proofing and many other visual control systems. However, the cleverest elements of the Japanese lean enterprise systems are not so readily visible to the naked eye – it is the process of policy deployment as we have come to term it in the West. This process was pioneered by Bridgestone Tire Company in 1968 and then developed by other Japanese manufacturers as a direct result of listening to Dr Edwards Deming preach his sermons regarding total quality management (TQM – a subject we will explore a little later in this book). It is the process of policy deployment that guides how the business and its many departments will focus on a few key 'step change' improvement programmes to satisfy customer needs profitably.

Policy deployment – originally called 'hoshin kanri' (which translates as 'directional control') is a system of management within which an annual policy challenge is set by the senior management team and then decomposed through the departments and organisational structure to a set of change tasks. The annual challenge itself is derived from a medium-term (three–five year) plan of what market and customer expectations will be in the future. In this respect, policy deployment seeks to predict the needs of the customer far enough ahead of today to ensure that the business is prepared for the new market conditions and can grow accordingly (Dimancescu *et al.*, 1996).

Central to the effective management of the annual challenge is the expression of the challenge in language that is shared by all managers. This language is typically that of the quality, delivery and cost of existing products and the time-to-market for new products. These aspects of business performance are shared by all managers and therefore they unite managers and highlight the 'dependencies' between them which we noted earlier. As these key processes pass through each and every department in the factory, the annual challenge is presented to the entire middle management group (department heads) rather than as in the traditional Western system of management-by-objectives where the managers are given improvement tasks which are designed to improve the efficiency of the department, regardless of the implications of these changes for other departments. Figure 3.3 shows how managers in different business departments are united by these core processes – the very processes which the customer wants to buy and sees value in.

By organising managers as a group, it is possible to use the group to understand and agree upon the best projects to improve the effectiveness of the firm. Organising in this manner therefore promotes Cross Functional Management (CFM), a holistic approach to company-wide improvement efforts, and gives all managers the opportunity to discuss best practices that are known to them (as well as lobbying for assistance and accommodation as they engage in departmental improvement activities). The engagement of managers in this way promotes dialogue and learning and, when focused on the 'annual challenge', this process results in the selection of the key improvement projects. In Japan, the negotiations of managers is often called 'catch-ball' and is a reference to the promotion of a change programme by one manager and the throwing of an imaginary ball to another manager in the room who will have to change their activities (Rich, 2002).

So let us put this into perspective, an annual challenge tends to be expressed as a percentage increase in quality and delivery levels with an associated percentage decrease in the total costs of making the product. Let us concentrate upon a 3% reduction in total cost – a manager may suggest implementing a lean 'Just-In-Time' production system and metaphorically toss the ball over to another manager who is regarded as

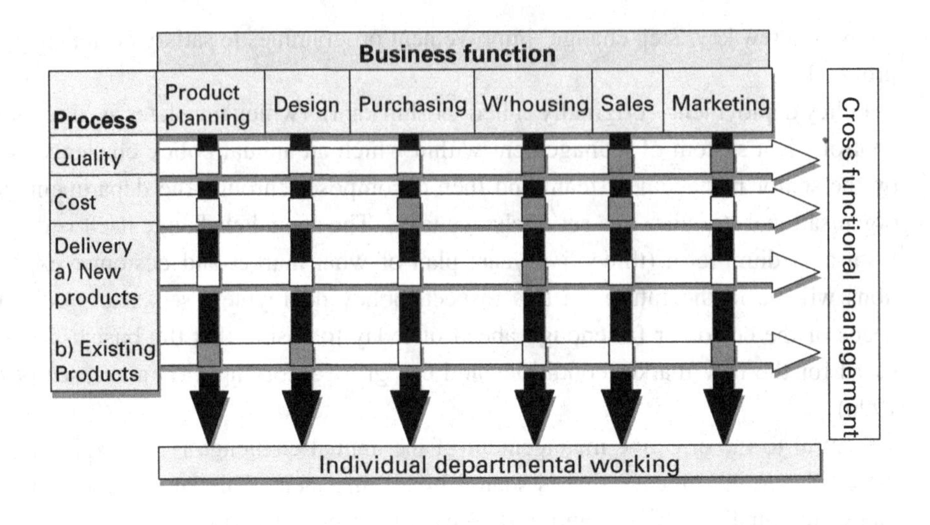

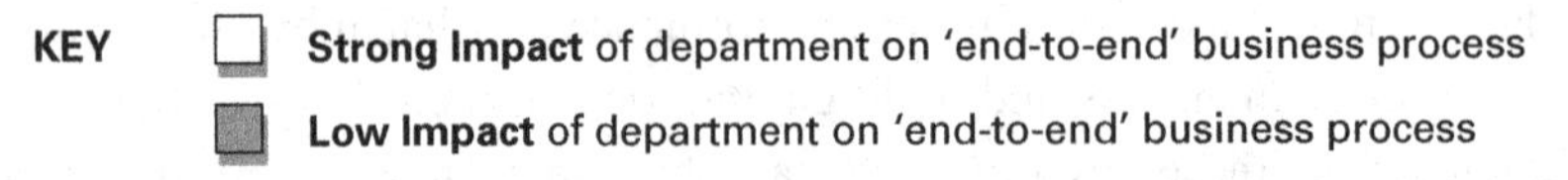

Figure 3.3 What customers want passes horizontally through departments

impeding the planning manager's ability to change and so on. An example could be the introduction of small lots and frequent deliveries from suppliers and the purchasing manager therefore becomes involved. And so on until the managers have agreed a course of action and returned to senior management to agree the way forward and to resource the projects.

In terms of timing, the senior management team will work upon setting the annual challenge a good four months before the end of the financial year. During this process they will spend a lot of time, as a group, learning about where customers expect to be in three years time. Around two months before the end of the financial year, the challenge will be set and the middle managers will be expected to begin 'catch-balling' and to resolve the process by the beginning of the financial year when budgets will be established for each project.

So at the end of the 'catch-ball' process, the key projects are related to the potential savings they could generate, the resources are assigned from all departments to support the programme and a senior manager is appointed to lead the change/review process. Throughout the new financial year, the key challenge indicators (quality, delivery and cost) are monitored through weekly project updates, which are fed back from the teams on a monthly basis at a meeting of all the middle managers. The monthly results are also fed back to the senior management, who will then compile a trend analysis for each quarter of operations. The quarterly formal reviews allow senior managers to understand the rate of change without having to be close to each project and therefore with the tendency to interfere rather than allow the teams to make their own decisions as management groups.

Near the close of the year, the senior management team reviews progress and sets the new challenge for the following year. In this manner, policy deployment keeps the business 'future focused' and reinforces the power of middle managers working as teams. This is the hidden guiding hand of the lean enterprise – a process which is well known to highly performing businesses as critical to becoming 'world class' because it is this process which makes sense of where and in what order to implement the well-known tools and techniques. It is a process about which you are well advised to read more and think through what it means for your business. If you return to our 'ideal type' lean business, at the beginning of this chapter, you will see how 'policy deployment' unlocks many of the management features of the lean enterprise, including deploying responsibility, cross-functional working, common measures and transparent information throughout the business.

Chapter summary

The essence of the lean enterprise is considerably different to the organisational practices most of us know and have grown up with inside our own factories. This chapter, providing a contrast and analysis of the modern lean enterprise, unlike most texts concerning lean management practices, has outlined the key design decisions of the senior managers in structuring and controlling the business. The chapter has provided the 'design logic' of lean businesses. It is this knowledge which is perhaps the most important to managers in understanding where the business is heading and the debates that are likely to be encountered along the way with other senior managers, subordinates, customers, suppliers and local trade union officials.

It has often been said that business managers get the culture they deserve. So if managers are confrontational they will generate a confrontational culture. Treat employees as fools and often the self-fulfilling prophecy is that they will behave like fools. Getting the culture you deserve – is not entirely true though, and most business managers have inherited the culture that was created by the previous senior management team. Business evolution and culture change are therefore time consuming and easily discredited if all senior managers do not act and behave 'as one'.

This chapter has provided a business-level analysis of the lean enterprise design and lean transformation process. The next chapter will put the change process into practice by exploring, at the operations level, the approach known as CANDO, which establishes the discipline of workplace organisation and control needed as a basis for your lean production system. In addition there are links to visual management, in

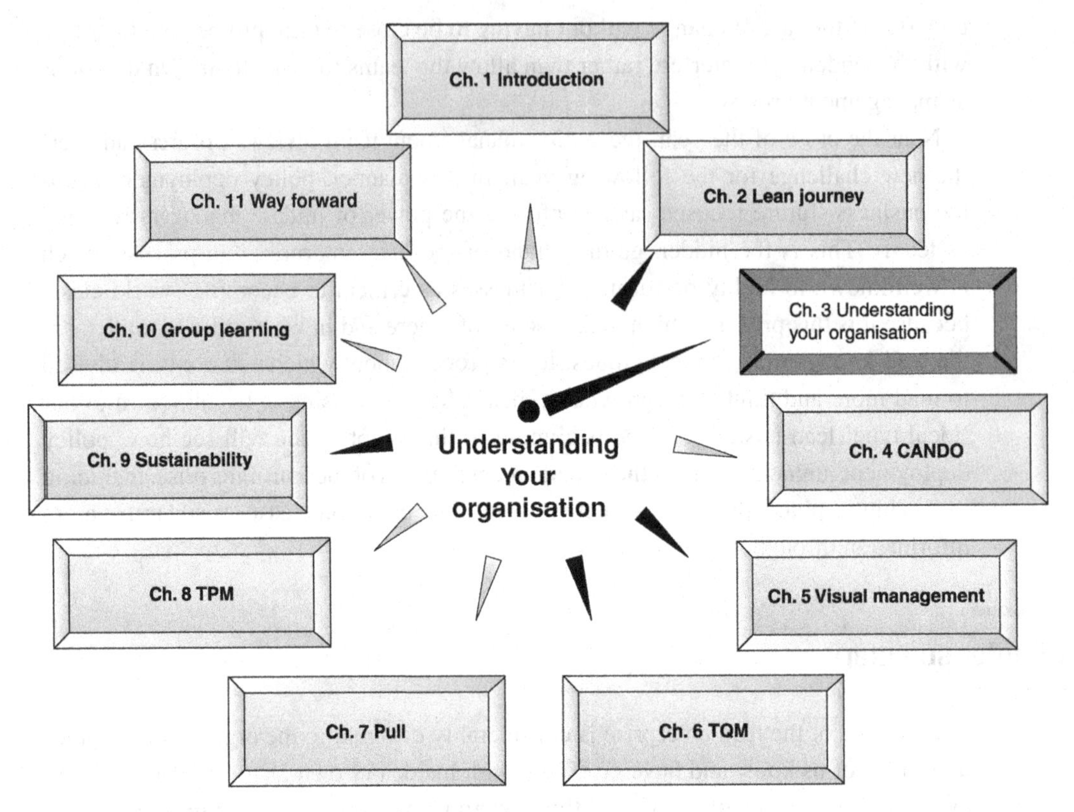

Figure 3.4 The overall production system features

terms of communicating much of the policy deployment and mapping outlined above, and TQM, as previously mentioned. The links to sustainability relate to the need for constancy of purpose and for implementers to understand where they are in the 'big picture'.

Signposts and recommended reading

The organisational aspects of lean production systems are far less detailed than the tools and techniques of lean production. But without the enterprise system and processes, the tools of lean production cannot be sustained and nurtured effectively. Cultures differ even between companies with the same technology and product mix and therefore the implementation process, content and speed will differ too. This chapter has attempted to summarise the key issues affecting management and their teams and, if you are eager to learn more, then Monden (1993) is a great place to start and an overview by one of the world's greatest Professors of lean systems[1]. Of course, Ohno (1988) is also worth reading as is Dimancescu *et al.* (1996) and Senge's (1993) publication on change management. For those of you who like techniques (those things that support but cannot replace a good management system) then try Rother and Shook (1998) or the toolbox of Bicheno (2000).

[1] See also Spear and Bowen (1999) paper in the Harvard Business Review (Sept–Oct 1999) entitled 'Decoding the DNA of the Toyota Production System'.

4 Laying the foundation stone of CANDO*

Nicola Bateman, Lynn Massey, and Nick Rich

Getting started

The first stage in the 'hands-on' element of the lean journey is to orientate people's thoughts and interest through awareness raising and promotional activities. The purpose of awareness-raising seminars is to provide key personnel with knowledge of what to expect from CANDO activities and to understand why these activities take place. The seminars consist of an outline of the principles of lean, a business game to illustrate lean principles and a review of CANDO activities. The lean business approach looks to change the system, but also to drive the change with the workforce by including them in the decision-making and change process. As a result, it is important that the whole organisation is aware of the lean business approach, and gradually the training in the application of lean principles will cascade down throughout the organisation and out into the supply chain (Dimancescu *et al.*, 1996).

CANDO is the most common starting point for the beginning of all change aimed at creating a lean production organisation and commences with the promotion of visible change within the workplace. Years of failed initiatives have resulted from good ideas that had little in them for those that were the subjects of the changes – the shop-floor teams. To create a positive management–employee relationship requires a clear signal from management that they are seriously focusing on lean as a business approach and not as a short-term initiative. The lean model illustrated in figure 4.1 shows the foundations for lean as workplace discipline and Control (CANDO). The CANDO programme, as we prefer to call this form of initial change, encourages workforce teams to critically evaluate the environments they are working in and to begin a process of improvement at their levels of operation. The purpose of starting here is quite easy to understand: it is visible, it is apolitical and it improves safety and improvement consciousness amongst all workers (Rich, 1999). In short, the tools and techniques of this stage favour everyone and include the individual, the team, the trade union and the management. The result is a clearer, cleaner, safer and more controlled workplace. This by itself is often enough to generate higher productivity, fewer defects, less accidents and the opportunity to build in greater flexibility.

* CANDO also known as 5S or 5C.

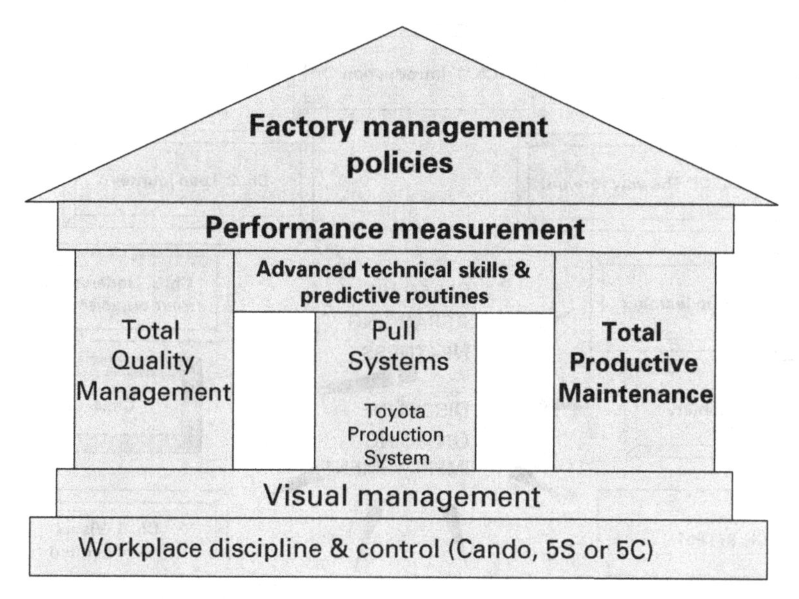

Figure 4.1 The lean model

This stage of lean implementation affects the safety and the morale of employees as they begin to apply basic problem-solving routines that will be repeated often throughout the remainder of the lean journey (see chapter 6). It can be difficult to 'sell' the idea of CANDO to fellow managers, partly because of its simplicity (often managers who are not familiar with operational areas think that many elements of CANDO are in place) and partly because it can be difficult to assess the financial benefits of CANDO as lack of workplace organisation hides problems and, until they are revealed by a CANDO activity, you do not know they are there. The way to sell CANDO is as the first stage to other change programmes: if you cannot get CANDO to stick, it is highly unlikely that any other change programmes will work, as they require the fundamentals of team involvement and commitment that CANDO can provide. As the CANDO programme evolves, it will uncover and identify new avenues on which to focus improvement within the workplace and more generally within the factory. These CANDO links are shown in figure 4.2.

A CANDO-controlled workplace lowers the risk of personal injury, it is an environment within which parts and materials are easily found and the CANDO'd workplace lowers the dependency of teams upon their managers. In this controlled working environment, teams make basic decisions and take charge of (i.e. the responsibility for) such routines as safety management, team building and improvement. For all these activities, the role of the frontline team leader is key as a motivator and also as policeman of the systems developed by the teams of employees that share the work area. As such, CANDO interfaces with many of the chapters we have put together to form this book and, it cannot be stressed enough, you must get this right and you must help your people

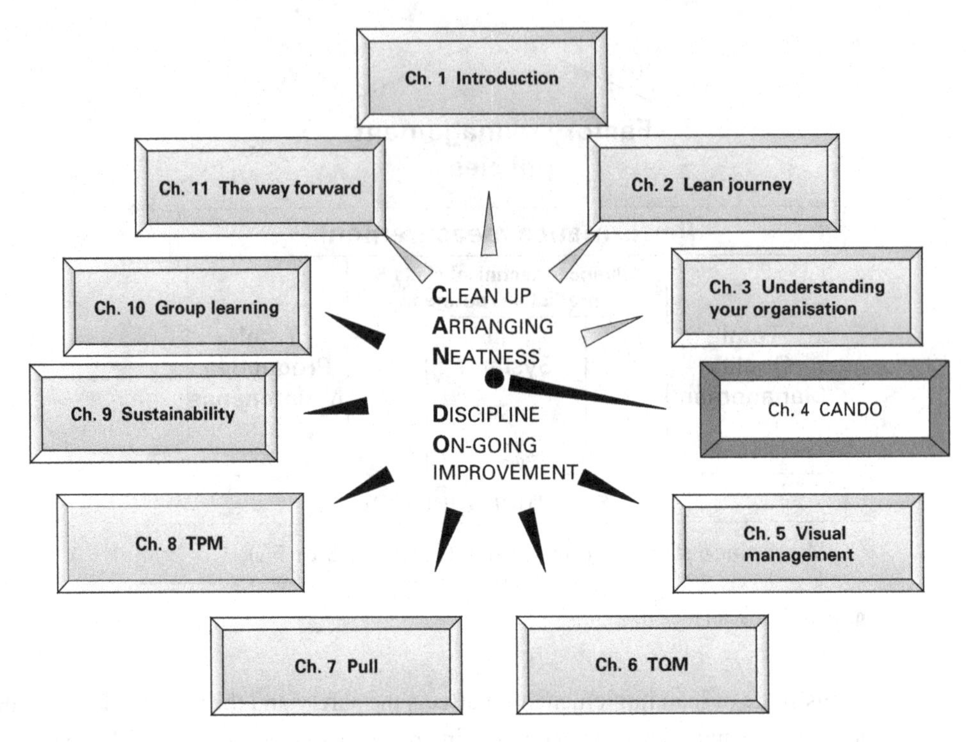

Figure 4.2 CANDO linkages

get it right. Fail at this most basic of disciplines and you will lose credibility, so it is important to spend time and as long as possible gaining the 'buy in' and ownership to creating a work environment that is also a show room to customers and other visitors to the factory. CANDO is a visual example of change and a new management–employee partnership; it needs time but it also needs to be promoted.

Box 4.1 Cosmetics and CANDO

In 2002, Cosmetics began its CANDO activities as a beginning to its lean journey of improvements. The programme commenced with the launch of a CANDO programme across the entire factory and its many different business unit teams. So successful was the initiative and the increase in worker morale that the factory was declared a 'show room' for the company and customers were encouraged to visit and participate in the improvement initiatives that resulted. The benefits achieved by the company, as a result of starting to build from the CANDO foundation, have included increased customer service levels, new business with existing and new customers, greater workplace involvement in decision making and higher levels of productivity. Also the company has become the subject of numerous local and national press articles about the factory and its ability to sustain the drive to 'world class manufacturing' status.

Table 4.1 *CANDO and its various forms*

Stage	CANDO	5S	5C
1	Clean up	Sort	Clear out
2	Arrange	Set in order	Clean and check
3	Neatness	Shine	Configure
4	Discipline	Standardise	Conformity
5	On-going improvement	Sustain	Custom and practise

The CANDO principles

This section starts with a brief description of the CANDO principles and then explores an example of how to structure a CANDO programme. CANDO stands for; clean-up, arranging, neatness, discipline and on-going improvement (see table 4.1). It is also known as the 5S's or 5C's. We prefer CANDO because it is an acronym that is easier to remember and better translated into English, but as long as the logic of CANDO is maintained you can call it what you like!

Stage 1 – Clean up or 'let's have a sort out'

This involves the work team sorting all the items that are necessary to carry out work in the production area and then to retain the tools, materials and documentation needed to support uninterrupted work flow. Objects are categorised by frequency of use and items that are frequently used are located near to the operator and items used less often are held nearby in the local area and infrequent items are sent to a centralised store for safekeeping. During the clean-up stage, some items will be disposed of or moved out of the area for assessment (these tend to be general waste, defective products and items that have accumulated over the years). Typically you will find all manner of defective and unused items and these can be 'tagged' (used to identify the item as 'not needed' and therefore 'flagged' for management attention and review) before being removed permanently from the workplace. All remaining items must therefore have a 'home' (the arrangement stage). It is often appropriate for area cleaning routines to be devised at the end of this stage (floor sweeping, area cleaning) so that the workplace is kept free of the debris that seems to collect in the factory and between shifts. These standards and materials are documented, assigned to workers and materials located close to the point of use. Also, by cleaning, the operators begin to 'police' their own factory areas and tend to become very aware about 'foreign items' left behind by fellow operators or dumped in the area by 'outsiders'. Another element of this stage is the establishment of workplace markings to denote the actual operating space for which the team is responsible. Everyone takes part in the improvement activity and everyone

● CLASS	1. Raw material ☐ 2. Work-in-progress ☐ 3. Finished product ☐	4. Machine \ Equip ☐ 5. Tools ☐ 6. Fixtures ☐	7. Other ☐
NAME			
QUANTITY			
SECTION			
REASON	1. Unnecessary ☐ 2. Defective ☐ 3. Non-Urgent ☐	4. Left over material ☐ 5. Other ☐	
ACTION	1. Eliminate ☐ 2. Store close by ☐ 3. Store in area ☐	4. Send to stores ☐ 5. Hold for analysis ☐ 6. Other ☐	
DATE	Tag attached : ___/___/___	Tag actioned : ___/___/___	
NUMBER :			

Figure 4.3 CANDO red tag (Rich, 1999)

'owns the problem' of how to keep the factory clean. Furthermore, they also share an interest in how to keep it clean with minimal daily effort (and minimal interruption to the available production time).

Hints for stage 1

Before starting to clean up, it is a good idea to walk through the area. Take pictures of the pilot area before you start the CANDO activities, as these help to identify problem areas and where to focus your efforts and resources. The pictures also form part of the history of the area and it is important that these are kept by the team. It is amazing the number of times you will hear that things have not changed much and these pictures are proof that the workplace has made radical changes from what was considered 'normal standards' in the past. An 'after' photo then gives a visual record to illustrate what has been achieved once the area is cleaned up. All items in the pilot area (tools, dies, consumables, inventory etc.) have to be sorted by frequency of use: items used all the time, frequently, occasionally, not used. The item's proximity to the operator is dependent on how often it is used. Items used all the time and frequently used items will be stored close to the operator and occasionally used items kept at a distance, with unused items having been removed from the area. When assessing the target area items, a process called 'red tagging' is carried out. The red tags (see figure 4.3) are attached to each item that requires an action to be carried out on it; this could mean removal from the area, relocation, disposal or repair etc. The tags can remain on the items for a set period of time to allow people in the work area to make comments (agree or disagree with action) regarding the action to be taken on the tags. If the item is to be removed, it can be quarantined for further assessment or for future disposal, relocation or repair.

Box 4.2 Clean up at Health Products

Health Products thought that they were very familiar with their workplace environment until they went out with a camera looking for problems such as: items wrongly stored or located, safety hazards, consumables poorly stored and allocated.

Learning points

- Look in drawers and closed cabinets to check contents and storage needs. You will find all manner of accumulated debris that is not needed and just sits around getting in the way of operators.
- Remember to look on top, under and above your normal eye level for possible causes of contamination in the area. You will be surprised at the amount of dust and other contaminants that build up over time.
- It is useful to have someone from a different area who does not know the area to carry out this activity as they will bring a 'fresh pair of eyes' to the workplace and will recognise problems more readily.

Stage 2 – Arranging or 'where should things go?'

At this stage we are looking at the ergonomics of the workplace and how the operators handle materials and tools and equipment. After the initial assessment, the area can be reorganised to allow easy location and access of all frequently used items (Hirano, 1996). This new arrangement should be easy to understand and maintain. The purpose of CANDO is to provide a work environment that is easy and safe to work in and is stable in terms of being organised in a known and consistent way. This provides an established, predictable work environment from which to make improvements. There is no point in trying to implement standard operations if every time an operator tries to reach for a tool it is missing or in the wrong place. Therefore everything must have a 'home' and this home is denoted by usage such that all items used within each shift remain close to the point of use (ideally within easy reach from the point of operation). All other items, say used weekly, should be stored in the area and lower usage items should be located in a safe store so that the item can easily be found when needed. At the end of this stage you should see the first visual improvement in the factory.

Box 4.3 Health Products and arranging the workplace

The company was operating in a congested environment that contained a high level of work in progress. The arranging activity helped the workforce to assess what was 'really' needed to be in their environment, with the result that a high proportion of the work area was cleared and they could set up their workplace more effectively.

Learning points

- Find simple and inexpensive resolutions to improving storage.
- Clearly indicate the walkways and areas to be kept clear – the operator can carry out their job without hindrance from people and materials.

Stage 3 – Neatness or 'getting things really clean and organised'

This stage starts with a large-scale clean up of the area, including all surfaces. The purpose of cleaning is to check in detail the elements of the workplace. Often at this stage further red tags are needed as minor problems such as worn parts or missing nuts are discovered. The neatness is then maintained by the team, with daily 5–10 min clean-up activities. Eventually this should involve everyone in the organisation. Further, the use of colour coding and marking techniques is extended to include tooling (shadow boards), cleaning materials held at specific locations and cleaning stations, consumable items (high and low markings) and naturally all safety equipment.

Box 4.4 Neatness stage at Health Products

The company already operated in a clean room environment and would be expected to have a high level of cleanliness. The operators undertook responsibility for cleaning spills and arranging their workplace from external cleaning contractors after finding that most cleaning jobs took only a matter of minutes per day. They introduced a colour-coded system for managing the cleaning materials and set their own stock holdings at each point-of-use. When materials got low they completed a simple laminated form and sent it to the central stores for replenishment. Also to avoid cleaning with the wrong type of liquid, a coloured sticker (corresponding to the liquid to be used) was fixed to a convenient and safe point on the machines itself. As such a green sticker denoted a solvent free cleaning liquid was to be used.

Learning points

- Think simple and try and bullet proof (mistake proof) the storage of all items that are held in the workplace.
- Use colour coding extensively, but ensure the colours used are common to all production areas.

- Always involve safety representatives and engineers to ensure all materials used and procedures are safe as well as efficient. Get the teams to document these procedures using a single laminated piece of A4 (with digital photographs showing what needs to be done) that is signed off by the engineering and safety department representatives.
- Train team members in the key technical and interpersonal skills needed to facilitate change.

Stage 4 – Discipline

This is the maintenance of stages 1–3. Standards of workplace organisation are established and each area will be audited to ensure compliance. The teams are responsible for addressing non-compliance, including carrying out continuous improvement and problem solving to rectify any issues. To 'police' this stage it is preferable to devise a rating score and to allocate the scoring into sections, including aisles, working areas, material storage, tooling and notice boards. A 1–5 rating (with 1 being no implementation to 5 being fully implemented) is the best form of this type of audit. You will also find that the teams will, by nature, devise audit score sheets that are very punishing, and where possible you should try to set up an audit sheet that can be used universally throughout the factory. The use of a universal audit helps in setting common standards and comparing overall team scores as well as targeting improvements throughout the factory to improve workplace organisation and discipline.

Discipline therefore means sticking to the systems developed in the previous stages. This stage cannot be done at a CANDO event and relates to time after the activity. This is when the hard work of having a lean work place starts. It is easy to generate enthusiasm for CANDO in a focused activity when people have the chance to organise their working environment, without everyday work pressures, but to keep up this standard with the usual hassles and problems of everyday working is just plain hard work. The chapter on sustainability elaborates on this theme, but it is important to stick to a plan, do, check, act (PDCA) cycle and follow up problems that are raised as part of the CANDO check list. This will often start to involve management higher up in the organisation, as CANDO will reveal underlying problems that need to be addressed at a middle and senior management level. Failure to systematically address problems highlighted by CANDO audits will undermine the whole system, so it can be useful to flag to your senior management the pivotal role that following up CANDO problems plays.

It is also important to close out all the red tags raised in the initial activity. A pragmatic approach is needed, but that should not prevent persistence in 'closing out' all the offending items that have been tagged. It is important that these 'follow-ups' are conducted, even if it is only at the rate of completing one item per week. Often there are a few persistent problems that are hard to solve, and it is important that the team gets to the root cause of these problems or they will not be tackling the fundamental issues that underlie problems within the work area. At later stages of CANDO development, it is likely that statistical tools and techniques will be needed to clearly understand and evaluate the problems in the workplace before robust solutions can be found.

Box 4.5 Promoting champions

Sean and Dennis both worked on the night shift at Medical Consumables as the CANDO activity began. Before long both had developed an interest in the programme and they co-wrote a training manual for all employees in the company (they were not asked to do this nor paid for this effort – they just did it!). This enthusiasm was duly recognised and rewarded as these two outstanding individuals were seconded full-time to CANDO promotion and training of their colleagues throughout the factory, warehouse, offices and wherever they could find an audience. Managers would be wise to note that this form of reward for 'good practice' and an 'improvement culture' goes down very well with the factory teams and it helps embed change.

Stage 5 – On-going improvement or 'Can we make it better?'

The final stage of CANDO is 'On-going improvement' and this means trying to improve the CANDO condition and the procedures that govern it. Managers should put pressure back on the teams to find ways of maintaining their disciplines in less time (therefore increasing the amount of productive time for the team). Encouraging teams to find ways to improve the system is important and will help to reinforce the learning (by repeating each stage periodically) as well as reducing the time and effort needed to sustain the CANDO system. Every minute released by improvement is another valuable minute that can be used more fruitfully, and, therefore, this is a form of continuous improvement as well as a periodic break away from the pressures of production. At this stage you will notice a change in the team to the point that they are now ready to begin solving other problems in the factory area and by default CANDO has taught most of the team a disciplined way of thinking about problems and their countermeasures.

It is also important to remember that the CANDO activities in the factory affect a number of performance indicators, including safety and productivity measures (see chapter 5). As such, it is important to display these measures and to identify why performance has improved and when initiatives were introduced. This helps workers and managers to see the impact of activities upon customer service and other performance measures. Such visual displays are important for later stages of the lean journey. Figure 4.4 shows one such measurement display that indicates how performance (measured by this company using an indicator called overall equipment effectiveness (OEE)) has risen. The OEE score (Nakajima, 1988), made up from the availability, quality and performance of the production line, shows a steady increase in the rolling average

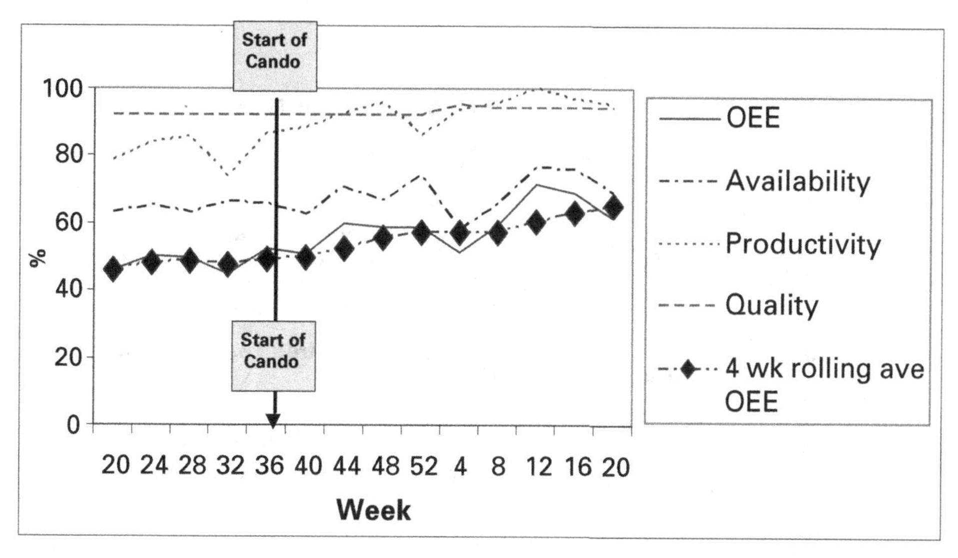

Figure 4.4 On-going improvement measurement

performance. Such a result is important and does help to reinforce the relationship between changes in working practices and improved performance.

Organising for CANDO

The previous chapters have highlighted the need for a change champion and change agent, and this applies to CANDO. This is the first real 'hands-on' lesson in lean production, and the individuals who spearhead and guide the implementation process must facilitate teams. Failure at this point will hold back your lean transformation, whereas the benefits of a firmly laid foundation will continue to generate benefits well into the future. The change agent's role is pivotal and the agent will need to organise teams of people for training and implementing CANDO activities.

In order for the team to concentrate their time and effort, the change agent has to be able to release them from their day-to-day jobs for the initial training and for the various 'follow ups' needed to 'close out' and realise the changes to the workplace. Further, the change agent must realise that a 'floating resource' of people may be needed to cover for team members' away training and ensure there is enough slack in the system to allow the remaining workforce to cover the absent team member's function. It is for this reason that we advocate a dedicated resource of a person or two to be devoted to CANDO promotion and day-to-day assistance with the teams and for these people to hold a budget to purchase items needed. Further, the change agent should control the CANDO review process such that activities and reviews are planned and decisions to support the programme can be made periodically.

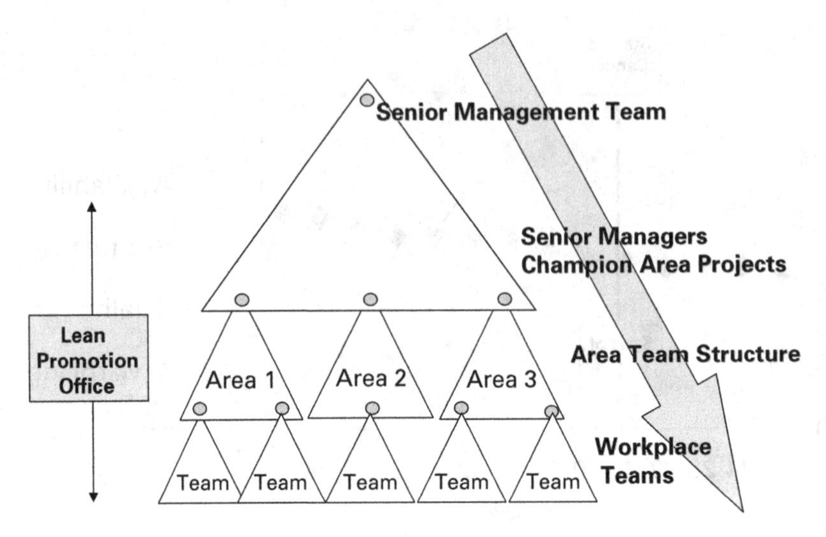

Figure 4.5 The Steering Committee structure

To shorten the time to bring about the CANDO transformation it is important to establish a committee structure around the change agent. This structure is shown in figure 4.5 and allows the correct deployment of responsibilities to the various factory teams and also a mechanism through which feedback on progress can be communicated regularly and in a formal manner.

Getting the 'doers to do'

CANDO success depends upon the skills and knowledge that can be transferred to the team. At the beginning, all the team is asked for is their commitment and support for the programme, so round one involves an initial promotion of the reason for, content of and likely activities involved in your CANDO programme. This is not a difficult task, as 'workplace organisation', especially safety in the workplace, are common wishes from teams and their union representation. Another facet of the CANDO approach is that it is a visual sign of management commitment to change. It is not a series of slogan posters on factory walls urging employees to 'work smarter not harder', but a visual commitment to training and the provision of the basic tools needed to maintain the workplace in showroom condition. This point is important, because, for the most part, business customers and even consumers are far more interested in visiting and auditing their suppliers than in previous years. It is the visual impression of the factory that is important in convincing these people that they should place their orders (and money) with your business. If your factory looks like a municipal dump, then this speaks volumes about the management and care taken in the factory towards productivity and quality. Poor workplace discipline and mess create the image that you are not in control, no matter how fancy a selling job you do with the customer! Put another way, it is like using CANDO to set up the factory as a showroom to sell more of your output.

So you may be asking yourself, 'What is in it for the doers?' This is a justifiable question. Basically the CANDO programme is a great introduction to a new world of problem solving at the operational team level and a meaningful way of interaction with the entire team. Such events add to the experience of working in the factory by allowing the individual to stand back and reflect upon 'why' and 'how' things are conducted at the factory. Furthermore it also allows individuals with key skills that have been gained outside of work to bring these skills into the workplace – such as those gained as a member of the Territorial Army or as a motorcycle enthusiast. Involvement in change therefore brings with it a sense of achievement, a visible indication of improvement and job satisfaction. These events are also social events and it is amazing that, even though workers may have been together for a long time, they may not know much of each other's lives outside the factory and about feelings when at work. However, in the most part, getting CANDO right is a first step in engaging in problem solving of a far more formalised nature and this increases the amount of training received and the option of studying for additional qualifications. It is therefore no real surprise that many companies elect to offer National Vocational Qualifications (NVQs) in such areas as business improvement, which include activities such as CANDO as part of the study course. So there is as much in the CANDO activity for the operations teams as there are gains for the business and customer service. The key to ensuring a smooth launch of a CANDO programme is to explore these issues with the change teams and find ways of adding these new motivators to enhance the sustainability of the programme and broader lean journey.

Box 4.6 Maintenance repair and overhaul facility

At an aerospace refurbishment factory, located in the UK, the teams decided to work on their CANDO improvements in small groups and to use co-workers from the offices to add a new 'pair of eyes' to the evaluation of the area's workplace management. This approach worked well and often these co-workers were more critical than the customers to the business. The net result was that the teams moved quickly from the classroom discussion concerning the reasons for change and techniques to implementation. These teams also deliberately used semi-permanent markings in the workplace (adhesive coloured tape) to denote aisles (yellow), storage locations (blue), calibration devices (white) and such like. This approach was taken to allow the team to modify and improve their initial improvements without using paint. The use of paint implied that when second generations of improvement were undertaken then the paint would need to be stripped-off, whereas the tape could simply be lifted up and the area changed quickly.

The CANDO team: inclusion not exclusion

The inclusion of key support function personnel such as production control, manufacturing management, technicians, quality, engineering and stores can make the difference between success or failure in setting out your new factory showroom. These individuals tend to contribute a good insight into potential solutions that the team might adopt. However, it is important that the operator teams and the frontline management lead the change process, with co-workers from indirect functions used as support. Also, if you are operating in a unionised environment, union representatives need to be part of the team too and you will find them the most willing of partners to this cause (and one of the most avid promoters of workplace management). As such, a cross-functional self-managed team approach, involving these various representatives, will make decision making faster and get buy-in at more levels of the organisation, as they have been part of the process.

A typical team of eight persons at 'Cosmetic', included representatives of the maintenance, quality and production control departments with the area teams. Each team also reported to a senior manager and the safety manager, through their area CANDO committee, and as such made presentations with their programme champion in front of the main factory board of directors and even in front of customers and suppliers to the business.

Box 4.7 Involving operators and the consensus approach

At a company we have worked with, the company adopted a 'technical approach' to improvement and devised a team that had the wrong mix of skills and allowed only one operator to attend the CANDO programme sessions (all other team members were technicians). This put pressure on the operator, who was unwilling to act as a representative for other operators. The initial meeting ended in the identification of a major problem of ownership and an admission by the only operator that, 'I am not going to tell them (other operators on his shift) what to do.' This is an important lesson and managers reading this chapter are advised to concentrate and lavish attention upon those people who work in the area and know the job. It is these people who need to be convinced that improved workplace organisation will lead to a better quality of working life. More importantly, it is these people who must be convinced to change their behaviour – so involve them. 'Exclusion' is a quick way to fuel factory 'gossip' and plays into the hands of the inevitable group of people who 'can't see the point in any of this'!

Learning points

- Must have operators on team.
- Make it clear that commitment to the event is required.
- Reduce poor time keeping and increase attendance by providing drinks and snacks (these breaks in the day allow discussions and learning points to be reflected upon – including what you would do differently next time).

By including support functions, there is a natural crossover into other business departments (an evolving form of CANDO that spreads to other departments) and also these dependents bring an alternative viewpoint that is very powerful when seeking solutions to team problems (including the elimination of the need to clean). As such it is important to be inclusive when conducting the workplace activities – everyone should feel a part of the programme and in effect everyone must take part if they are then to audit or detect slippage in factory performance. At the end of the CANDO activities, the output is all about resetting employees' standards of what is an acceptable condition for the factory and you can never have enough people setting high standards.

Box 4.8 Medical Consumables and the MD's office

Medical Consumables is a 'world class' manufacturer of sophisticated products (winning many awards for quality and productivity). The MD established his CANDO activities in the factory and these rapidly spread to other parts of his organisation. An example of this was the application of the approach by his two personal assistants, who organised everything from paper and photocopier supplies to cleaning routines for the fax machines!

Learning points

- Practice what you preach – 'staff' or 'indirect' operations are not immune to workplace discipline – they are workplaces!
- Shop floor teams should be encouraged to audit the indirect areas of the factory in order to prevent demarcation.

Selecting the pilot area

It is a good idea for the change agent to have a particular area for the programme to start, as this will allow him/her to assess the amount of work that will need to be done and the resources that may be required. With this in mind, the change champion and agent can guide/make suggestions to the team and the team can then influence the 'success' of this pilot activity. Here we offer some basic pointers to consider during the CANDO foundation activities.

CANDO activities

These activities are used to promote communication and relationships between management, employees and functional areas, bringing together cross-functional teams to train together and implement the CANDO activities. This is an important element in the lean journey. During the training, there are opportunities for the team members to question how things are done and to begin to understand where they fit into the total value chain. The training activities begin to level out some of the tacit knowledge that is held in the organisation. A stronger relationship is developed through the sharing and exchanging of thoughts and ideas about the current system and debate on ways they could improve it in the future.

- Pick a very visual area of the factory (right in the middle or an area that is passed by all employees daily as they head to their individual work stations). This approach increases interest in the programme and often results in approaches to the steering committee and change agent from other areas (increasing internal demand and a pull from the workplace itself rather than coercion of employees to buy into improvement).
- Avoid 'bottlenecks'. It is tempting to select the bottleneck operation and go for a 'quick hit', but this is often regretted. The reason why these are to be avoided is simple. Managers do not like to release these assets from production (and get annoyed when you repetitively ask for time on these machines). So the worst thing you can do is to pick an area to which you will have to fight for access. The starting point for CANDO is to establish a 'showcase' from which others can learn and for you to establish the right toolbox for your own training and development activities.
- Don't take on a huge area – small should be made beautiful. Taking too large a portion of the factory creates a lot of administration problems and increases the co-ordination requirements of the programme. This will lead to a dilution of the learning points. It is far better to start small and create a showcase to which you can bring others to see 'good manufacturing practice' than to start by taking on too much and having bits and pieces of good practice spread too widely.
- Before launching CANDO have a think about the taping of aisles and storage locations for items such as raw materials, work-in-process, finished goods, cleaning materials, calibration devices etc. By setting the colours that denote these items before you start will ensure a commonality of approach throughout the factory (and help you to buy the right colours of tape). We recommend you use tape, where safe to do so, for such floor markings for another simple reason – it can be lifted back up. CANDO activities involve constant change and re-allocation of space, so if you paint the

floor you will face a costly bill when you need to sandblast the paint off and repaint new areas.

Box 4.9 Mornington Cereals

Mornington Cereals is a large-scale and continuous processer making breakfast cereals, and the main packaging area was chosen as the pilot area because:

- Processes up-stream from packaging were due to increase capacity (from a considerable capital investment as these assets were the bottlenecks to the factory) and so the packaging area would need to be able to keep pace. Any efficiency improvements at the packaging area would therefore be translated directly into customer service improvements.
- There was little capital investment planned in packaging, so unlike other areas of the plant it would be possible to get access to the area.
- The packaging area was cramped and needed to release vital working space (seen as very important by the factory teams).

The main production line within packaging was selected because it was the highest volume line and so a capacity improvement of the main production line would have the greatest increase in capacity for the whole packaging area.

- Engage the safety and maintenance/engineering functions with the CANDO change programme at the steering committee level. As described earlier, the links to maintenance and quality are pivotal. The standardisation of documents can directly impact on quality audit trails. The 'Shine' and neatness stages link directly to the broader issue of team-based autonomous maintenance and also to preventative maintenance routines. Departments such as maintenance engineering will regard CANDO as a foundation upon which they can build new working practices and a programme that will give them benefits from participation.

At the pilot area level, you will need to consider the following points:

- **Talk to the team about a good place to pilot the programme.** From the start involve the team in the decision-making process to get their 'buy-in' into the changes that are going to happen.
- **Allocate somewhere to quarantine and if required dispose of items.** The team will move items out of the area that they consider unnecessary. These items can be placed in a quarantine area for assessment for retention or disposal.
- **Ensure those not in the pilot team know what is happening to the workplace.** Respect that people may be reluctant to change their environment, especially those not directly bringing the change about.
- **Plan how the pilot will be rolled out.** Think about who and how you will resource the rollout. Consider how to include everyone and bring people from the pilot area into the new teams to help them learn how best to engage CANDO.

Box 4.10 Mornington Cereals (too much talking and not enough action)

Reflecting upon the pilot stage of the CANDO activity with the team, the change agent and team decided that too much time had been spent on discussion and not enough on hands-on improvements. The management team decided the solution was to run some awareness raising days for key personnel. This would enable people to know what to expect of a CANDO activity and understand how CANDO fitted into the larger lean transformation.

In subsequent activities it was clear that some of the fundamental issues about how the factory was to run had been clarified and people had a clearer idea of what was required of them.

The CANDO roll-out process

The CANDO process should be extended beyond the factory floor to other parts of the business, such as administration areas, sales or stores. Having created a 'show case' and working example with which to show people in the factory the concepts in action, it is possible to train virtually all other employees and make benefits in support functions and even with suppliers/customers. Within the 2–3 day initial training, the team should be able to reach stage 2 and have a list of activities to take away with them for stages 3 and 4. Stage 5 will involve further training in tools and techniques to support them in their improvement activities.

The second part of rolling out CANDO is focusing on the discipline and on-going improvements. The practical training behind CANDO can be delivered in a short period of time, but the art of good workplace organisation is practice and sustaining the achievements made not just for a few weeks after 'kick off' but by making it daily practice (especially for new employees to the area). In our experience CANDO programmes are highly enjoyable and they rarely fail to make a positive impression on the workplace and employees throughout the factory. To get embedded in 'customary practice' however, requires application and continuous improvement. If CANDO starts to slip then it is probably a sign that morale in the factory is falling and the CANDO system needs re-invigorating. Or it may mean that new personnel in the area need training. As such, CANDO never ends – it just takes on a new form or evolves into a new set of activities that eliminate wastes in other aspects of what is done in the workplace. Most of all have fun with this element of lean production. System implementation after all is the foundation for greater things to come and for initiatives with much greater business benefits. Get the foundation wrong, or turn it into a boring and bureaucratic exercise, and you will have achieved nothing, slowed down your implementation plan and 'turned off' those co-workers who initially gave you the benefit of the doubt, but now believe its just another initiative to make people do more for no reward.

Reflections on CANDO

We have worked with a very wide range of companies, from large to small and from the jobbing shop to highly automated factories – each experience adds a new dimension and set of learning points. At each company these CANDO activities have included manufacturing, administration, quality documentation and materials preparation personnel and there are many common lessons learnt from these experiences that will provide some insight for individuals looking to launch their initial CANDO activities. Here are some tips and hints:

Avoid over-theorising about CANDO – it is not a philosophy, but a way of working. It is better to gain credibility as a manager by starting the process and gaining experience rather than wasting time deliberating all the imponderables you might encounter.

Get the advice of others (visit others) to get an understanding of how things could be improved cheaply and how to organise effectively. This is particularly true of the safety representatives at the factory and in ensuring there is total compliance with safety rules within the factory and that all involved in CANDO understand/comply with the requirements of the factory. It is hard for managers who have been in the job for a long time to suddenly expect change from their co-workers (all manner of baggage gets in the way), so talk to people and understand their concerns.

You should talk to your resident trade union representative if you have one (these people will tend to tell you straight what needs to be done). Promote the personal and business benefits of CANDO. At the end of the day you must have a purpose to start CANDO. At the most basic level it is about safety and productivity, but these have impacts upon how the business sells itself and what future activities CANDO will help start. Even cynics will play a role just to see where it leads. Be careful to ensure co-workers attend the CANDO programme meetings, but also be sensitive, when planning, to issues such as childcare problems, inevitable factory distractions and ensuring people do not leave the implementation but are focused on improvements. Finally, ensure that everyone knows this change programme requires participation, including physically being involved in work place change.

Avoid looking for cost savings and be willing to spend money on helping factory workers to help themselves. Senior managers who demand 'pay back' are thinking too narrowly – CANDO is not about payback in two months, it is about changing attitudes and increasing morale so payback does not come into the equation. If you cannot get CANDO to work, then you are certainly not going to get your lean production system to stick – so forget it. Concentrate on delivering success in whatever form.

Do not spend too much time on presentations – learn by doing and always get the team to reflect on what they have learned. We do not spend enough time listening to our teams, letting them talk. Computerised presentations, with fancy slide transitions and animation, is not what CANDO is about – good quality but minimal classroom training and maximum doing are the keys to success.

Show people a video of the factory and pictures of offensive items/areas. This is a great 'wake-up' call as co-workers adopt the role of customers and watch a video tour of their own factory. It is surprising how decades of 'that is how life in this factory is' can be changed as co-workers see the factory as an outsider or imagine taking their family for a tour of where they work. Images are great ways to learn and work well – you should use them to give reality to CANDO and win support for the initiative.

Box 4.11 Views from the Learn 2 companies

Here's some feedback from CANDO Teams with quotes taken directly from workers we have spent time with

'Until you actually do it you think it [the workplace] is OK – it opens your eyes.' (Operator)

'There was a lot of criticism . . . you just have to learn from that and find solutions with people.' (Operator)

'We can see the difference! . . . its going to make the job easier and other people will benefit too.' (Operator)

'Our members have thoroughly enjoyed their experience of CANDO and it has brought a new dimension to our health and safety meetings and provided a mechanism through which some of our issues can now be managed autonomously and countermeasures devised with the teams.' (Shop Steward)

Views from management

'CANDO we thought was important in terms of a foundation, it is also important in terms of people, behaviour and the environment.'

'We weren't convinced that starting with CANDO would give us anything . . . I guess we just took a punt and had a go . . . it paid off for sure!'

'We can't tell you exactly how productivity increased but we know that CANDO was the catalyst and for the first time in our history we had employees asking to participate in change. In all my years as a manager I've never had people moaning that we were dragging our feet in getting them involved. It was an eye opening experience and the more responsibility we gave to the teams, the more they wanted.'

'If we were to do it all over again, I'd give every team a budget for materials. We found that the teams, when they had money, were the meanest at spending it and naturally looked for ways of maximising what they could get for every penny they spent. We even had one group who went to B&Q one lunchtime to buy bits and bobs because they were cheaper than our quotes.'

Conclusions

For many companies this will be the first adventure in lean thinking and it is one of the most enjoyable stages for the entire workforce as the system of workplace organisation is introduced. The key thing to remember is that the business is 'on a mission' to convert the workplace from its 'current state' to a controlled and disciplined environment and to reinforce the lean approach of getting change executed by integrating the people who do the job. As such, it is important to avoid blame – do not focus on 'how did we get here' but rather 'how do we get to where we want to be'. The benefits of the CANDO system and the achievements of the teams must be praised highly to encourage greater and greater levels of participation. It is important that the key processes of 'safety' and 'morale' of factory employees are improved so that individuals, even cynical employees, see visual change and benefits of the new way of working.

Signposts and recommended reading

Whenever we start a lean transformation programme with a business we get involved with a CANDO activity. The intensity and enjoyment of this change activity is always impressive. There are a few books written about the subject which are worth reading namely Hirano (1996) and Shirose (1996). A couple of final signposts for you include getting in touch with your local office or website of the Department of Trade and Industry (DTI). For small businesses also see the Manufacturing Advisory Service (MAS) in your region for local support.

5 Visual management and performance measurement

Nicola Bateman and Nick Rich

Introduction

The 'art' of designing a fully effective lean production system requires attention to detail and a keen eye on the standardisation of processes in a manner that makes working life simpler and better for those who work in the factory (Grief, 1991). On first encounter with a system that has been implemented with full visual management practices, observers find it a contradiction, the almost-clinical approach to the design of the workplace seems to conflict with the high premium lean manufacturers place upon team autonomy. In reality, there is no contradiction because visual management creates a standard which allows 'abnormality', in whatever form, to be detected quickly and frees the worker from thinking about the basics of good management in favour of applying thought processes to more important matters of improvement. It looks simple in operation but visual management is an art in itself and a lesson that is well worth learning as it provides a path for problem-solving and improvement activities in a way that can be sustained due to its simplicity and 'ease of doing' on a daily basis.

Examples of good visual management surround us all – the red markings on the rev counter on the dashboard of your car and the level markings on the side of your kettle to avoid overfilling, and even the different colours of fire extinguishers are just a few. If you look around you will find many more which you have taken for granted but implicitly understood and more often than not conformed to – like parking in a designated space or not parking where double yellow lines have been painted on the tarmac. These symbols, signs and colours help us to adhere to rules, without which there would be chaos, and the same is true of the workplace. The purpose of this chapter is to build upon the first steps of the earlier chapters to provide an overview of visual management as the basis for workplace improvement and to introduce shop-floor-based performance measures. These visually displayed performance indicators link improvement activity with positive changes in key measures as well as showing when remedial actions are needed to bring the performance of the team back to the standard expected by the organisation.

Area \ Purpose	Manage	Improve
Operational	• Maintain current condition	• Provides known standard from which improvements can be made
Communications board	• States standard e.g. 5S check list	• Allows informed improvement decisions to be made

Figure 5.1 Role of visual management

Visual management

Visual management (VM) is one of the most versatile aspects of lean production systems and yet is also one of the most neglected elements of workplace management. Visual management is about making the workplace more visual and this approach covers all aspects of work and the workplace. It includes performance measures visually represented in graphs on walls instead of hidden away in desk drawers, having tools and equipment configured visually so you can see if they are missing and having work flow that is simple and obvious.

VM is particularly versatile because it can be applied to almost any work situation and has an application in every factory from the manufacture of metal pressings to the control of nuclear power plants. Visual management is not limited to repetitive work environments like some other elements of the lean approach such as standard operations to control the work tasks to manufacture high volume parts. As such there is no reason why a 'creative' work environment such as a design studio should not employ the principles of visual management. Indeed, you will find visual management practices in television studios, hospitals, aircraft refurbishment facilities, which operate with low production volumes, and in circumstances where no two days are the same.

Visual management on the shop floor has two main areas of application (figure 5.1). The first application concerns the control of the operational area and working environment, for example in the form of tool configuration (such as 'shadow boards'), and the second in reporting facts and figures about the work area in the form of the communication board. Both of these uses of VM serve to manage the workplace by maintaining the current conditions set out by the team and factory policy but also as a foundation for improvement of these standards (Grief, 1991). Both 'total quality' and its influence on the development of the lean approach stress the need for standards and the use of such standards to form a basis for improvement activity (a ratchet effect). This legacy of Deming (Walton, 1992) means that improvements become integrated in the form of standard operations and VM such that they form the basis of the next generation of improvements, before these then form the foundation for the subsequent generation of improvements and so on.

The role of VM is therefore to act as a ratchet and to assist operators in maintaining the current condition of the work area by identifying any abnormalities and deviations from the accepted standard and to prompt the team into taking action (Hirano, 1996). Also, even if an abnormality does not occur, then these standards allow teams to take proactive improvement activities.

There is one further addition to VM which has a great payback to the firm. This comes from the increasing trend to recruit temporary workers and also hopefully for the business to recruit new workers. For these new employees, the factory is a confusing environment of product flows and there is a lot of information to take on board before the worker can be an integral part of the team. To allow these people to understand, learn the process and to make good decisions about what to do, visual management is indispensable. Shortly after the Learn 2 programme had finished we visited a national distribution centre of a large UK supermarket *three weeks before Christmas*. As you can imagine, this is the food retail business's busiest time and they employ many temps to keep up with demand. Despite this, the distribution centre had a sense of purpose, although very hectic, people knew what to do and where they should be. What struck us was that the simple VM employed reduced safety issues – there were many large forklifts whizzing around – and directed the less experienced staff about their work. Just think about the experience you faced when learning to drive – with little experience of driving, you tend to over-rev the engine, but, when you are instructed that 'red lining' is not a good way to drive a car, then whatever vehicle you get to drive, you tend to know what is abnormal from just this one simple mechanism. With this simple lesson in driving, you can then progress to take on more information and so your learning potential and capability expands. To get new workers 'up to speed' quickly is a major competitive advantage as well as a means of preventing costly mistakes. This is not a traditional means of inducting new workers. The tradition was to 'sit with Nelly' and learn by watching someone else and, in so doing, picking up their bad habits without questioning whether these were safe or indeed efficient.

To assist existing and new workers, operational VM includes tool shadow boards, painted placement squares (figure 5.2) and colour coded equipment inspection points. The placement square in figure 5.2 clearly shows where boxes should be placed and how many boxes can be placed in that square. Imagine if these boxes contained heavy die-castings of engine blocks and it was your responsibility as a new worker to stack these incoming materials. Unless you have been told that stacks over three boxes high are dangerous or read the large materials handling manuals to find the one instruction (handling of die-castings at Goods Inwards) you would not know what is a safe stack height and what is not until the point at which you get it wrong and either receive criticism from another employee (usually a manager) or an incident occurs which could have far more devastating consequences for those involved. Many industrial incidents,

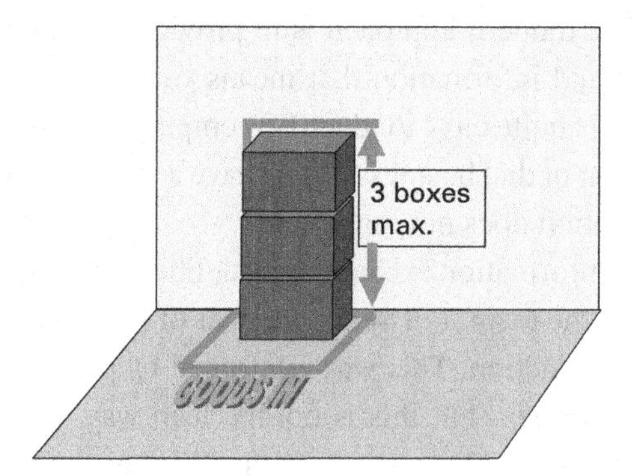

Figure 5.2 Placement square

some fatal, have occurred in environments that were suitable for VM and for which VM could have been an effective countermeasure to 'getting things wrong'. The simplicity of VM routines and the attention to detail needed to ensure that work processes are controlled in this manner is not quite as 'simple' as many people first believe. If it was that simple, then perhaps more team leaders and managers would promote its application to every part of the factory. Surprisingly, many managers will spend a huge amount of time writing policies and procedures, yet it is VM that, at the front line of operations, provides the most useful indicator of what is acceptable and what is not. You do not have to read and write to understand VM – it is almost a universal language whereby you can understand the condition and standards expected.

To this extent airports are designed with such features in mind because of the different people speaking different languages that pass through them. Just imagine, if you turned up at Heathrow airport and everything you needed to know was written in a procedure manual or worse still it was written in the same format as a typical quality manual as in most Western factories, you would never catch a flight and Heathrow would be a sea of passengers walking and reading at the same time. So VM is instrumental to teaching, learning and conducting tasks in the factory in a manner that is standardised and to an acceptable standard

Communications boards

For years and generations of workers in traditionally managed factories, information was hidden, often deliberately. The traditional logic was to prevent everyone from having too much information which could be used to highlight the weaknesses of management. Thankfully factories that work in this manner are getting fewer and are being replaced with a system of 'open information' about the business, the performance of the team to which you belong and also about customer issues concerning service levels and product information. This transparency reinforces a new belief in utilising all workers' knowledge, experience and ideas. The modern approach is to provide information in a way that everyone can understand and information that means something to every worker (Grief, 1991). The motivation is quite easy to identify – employees can make a greater contribution to the improvement of that business if they have a wider knowledge of the organisation. So hiding information does not make sense.

There are many benefits of opening information exchange and deliberately displaying information about the performance of the team. For years, the fear of punishment was associated with not making enough production. This was reinforced by piece rate pay and the drive to make as much as possible – but this is not the lean way. You make only what is needed and no more – that would be a waste. In the old way, the numbers reported from each production area would often be manipulated in a skilful game of reporting numbers that management wanted to see. These games did nothing for any employee of the factory – they were kidding themselves and providing management with untruthful information upon which the management would make decisions. All this deceit got in the way of managing and was a product of the way in which the factory was managed (Kolb, 2000).

It is no surprise that Deming (the greatest of all total quality management gurus) once called for the 'elimination of fear' in the way supervisors reported performance information to factory management and also the stopping of 'the quota mentality'. His point was simple, get rid of these two issues and managers will receive open and honest feedback on the performance of the production area. This honest reporting is needed to prompt improvement and problem-solving activities. In the past there was very little problem solving because this was the responsibility of a 'support department' such as engineering or quality assurance, but now employees need to see the real numbers if they are to make a swift and effective response to deviations in performance from whatever source.

Critical to overcoming the 'reporting problems' and the issues of creating a standardised method of reporting is the introduction of team communication boards located in the production areas and providing meaningful trend and target information so all workers can see how well, as a team and not just the leader, they are performing as an autonomous part of the factory system or value stream. If you like, it is a means of establishing the performance standards to support the flow of materials. These visual standards therefore serve the same purposes as the physical ones reviewed at the beginning of the chapter – to monitor the status of a team and to identify areas for commercial improvements to the efficiency and effectiveness of the team. The measures that are displayed upon these boards are therefore key indicators of team performance and monitors that help the team to learn about the process and its improvements (Grief, 1991; Rich, 1999).

Some academics have criticised such displays of information as a means of 'policing' teams and using numbers to instil a sense of fear, but this is not the case. It is pretty much a right for people to get feedback on how well they are doing – the basic building block of a lean organisation is its teams, so this is a means of feedback to the team. The traditional alternative was for a manager to enter the factory and rant at the workers to improve productivity, but this system was not effective because the teams did not know, from one week to the next, what the expected standard of performance was set at. In reality, the communication board is a control mechanism, but one which is there to enlist involvement and engagement of the team in a more sophisticated manner than the traditional 'telling off' approach. It is also there to show and attract praise for the improvement efforts that have taken place. Finally, the board acts as a reminder to employees that 'things have changed' and 'for the better' as these boards show the areas of improvements and inform on projects that have been 'closed out' as teams begin to take on a true self-management responsibility.

Communication boards show items that should be monitored by the team and, in the interests of engaging teams, in learning about the processes they control and shortening the time between detecting a drift in performance and solving the cause of the drift, the communication boards are there to prompt action. This is much the same as task-related visual management techniques. They should be used by team leaders as part of the start (or end) to a shift whereby the team leader briefs the team as to the day's activities and runs through any issues that are occurring. This time can be used for team decision making. Initially the team leader may find this time consuming, but as they and their team become more practised the brief will usually take about ten minutes. It can be useful for team leaders when initiating this process to have a 'crib' list of topics to cover.

Communication board measures

There are many forms of communication board and to some extent each board is tailored to meet the information needs of the local team and the 'in-house' factory style. They range from large magnetic boards the size of a pool table to small boards mounted alongside machinery.

There is a distinct lean logic to the information provided on the boards which is rarely stated but it does have an impact on what information you give to teams and why it is important. The measures selected for the board need to be carefully chosen, as they should guide improvements and actions in the cell/production area. Too many measures can be overwhelming and as uninformative as none at all. And the combined team measures should provide a narrative (or story board) explaining events and actions. The main categories of information provided by the board and the associated logic of presenting this information are as follows (see figure 5.3):

1. Information about the morale of the team – there is broadly an acceptance that a good factory has workers who are, on the whole, happy to come to work. So these measures will focus on the key control item of 'attendance'. Many people will have

been used to measuring 'absenteeism' but 'attendance' is a positive measure for which it is better to record a 98% achievement than a 2% loss. It is psychological, but the more positive the approach and the higher the measure the better, when it comes to assessing people rather than processes. Another measure recorded under the heading of 'morale' is the number of 'improvement suggestions made' or 'projects completed' by the team. These show whether the team is actively engaged in problem solving and is integrating all team members in rooting out waste.

2. The second bank of information is displayed under a title of 'safety' and this is another critical control item which is often recorded as the 'time between reported accidents'. Obviously a people measure and the longer the time elapsed between incidents is another sign of positive improvement. This measure is also a key measure for trade unions and shows whether people are being safe at work. If 'safety' management starts to slip and incidents or near misses occur, then it is time to address these issues – probably by revisiting any slippage in the CANDO workplace organisation and procedures.

With these two sets of measures in place, it is easy to see whether team members are coming to work, whether the workplace is safe (the basic right of any worker) and whether people are engaged in looking for improvement. Basically, if a team does not fare well on these indicators, then there may well be some real issues concerning the ability of the workers to act as a team. These are the basic measures and teams that perform well on these measures are likely to exhibit the discipline and approach needed to make good progress at the next level of measures which monitor the status of the process and its management.

3. The first major group of process performance measures concerns the 'quality' performance of the team. This is the 'number one' process-based priority and is used to track the 'right first time' performance measure. This measure is important because it tells the team how well they are at working without any physical waste or how well the team do at avoiding rework situations. If the quality performance of the team improves then this will lead to lower costs and better productivity.
4. The next set of measures concerns the 'delivery' of production outputs in a timely manner and either monitors the 'schedule adherence' and/or the status of the pull system and availability of materials. If quality is good and the delivery performance is good, then this will help reduce costs further and result in a good flow of materials to the internal or external customer.
5. Finally there is a measure of 'cost' which tracks items within the manufacturing process, which should be improved to reduce the costs of operating the production area. These costs are typically the costs of waste in the production area, such as the cost of defects, the excessive costs of lost production due to long set ups, the costs of overtime working due to rework of defective items. These controls do not relate to the costs of people, but the costs of poor process performance. This is the final

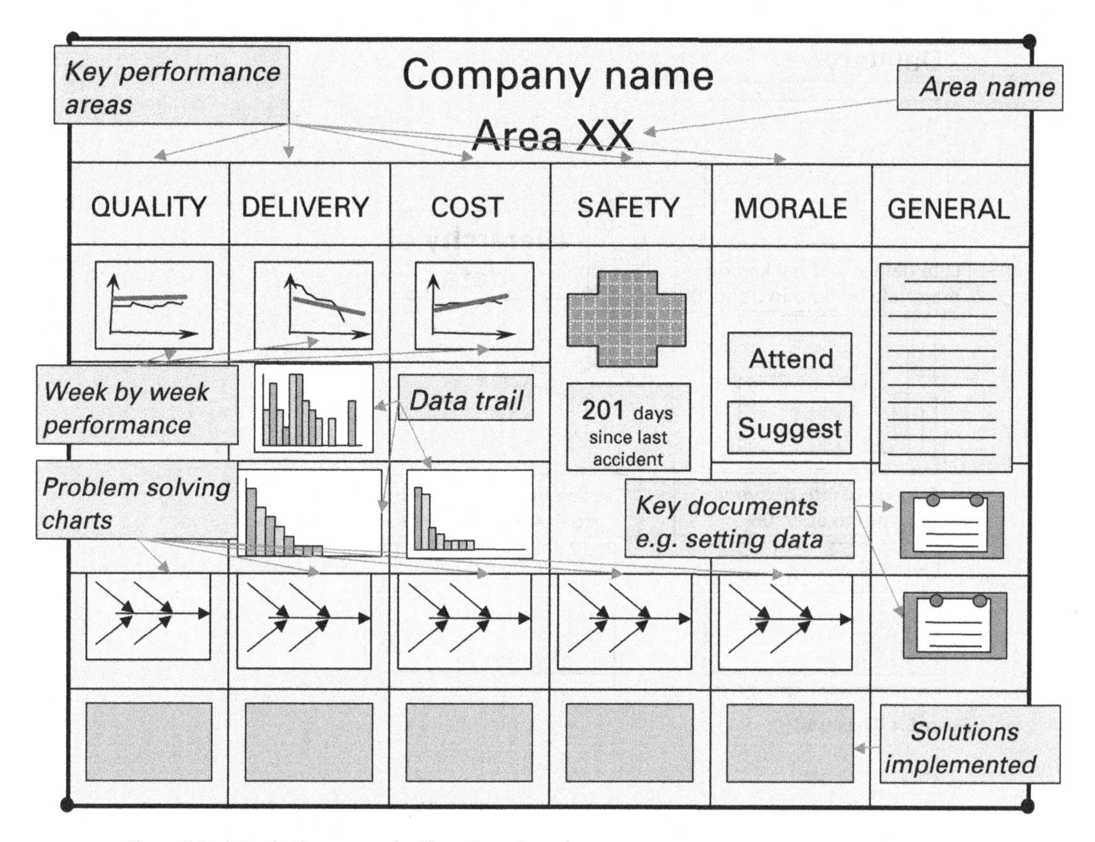

Figure 5.3 A typical communications board

target for process improvement after the quality and delivery performance of the team have been secured. It is the last point of improvement, whereas traditionally this is the main performance measure for traditional businesses and it is this focus on 'cost-first' that causes so much trouble for mass producers.

The logic of mastering the business processes' quality and delivery before attempting to take away all the additional unnecessary costs is based on getting the quality right which is a prerequisite of getting the delivery right which is a prerequisite of reducing costs (Rich, 1999). Such an approach allows team members to practise and employ the problem-solving techniques necessary to reduce costs in a controlled manner. The combination and presentation of these measures to teams is a powerful way of getting team members interested in performance and improvement issues. Other measures you may find presented on the communications board include general business notices (pensions, union notices etc.), measures of team productivity, and also, especially for companies engaged in total productive maintenance (TPM), the overall equipment effectiveness (OEE) measurement. The OEE measure tends to be shown as an indicator

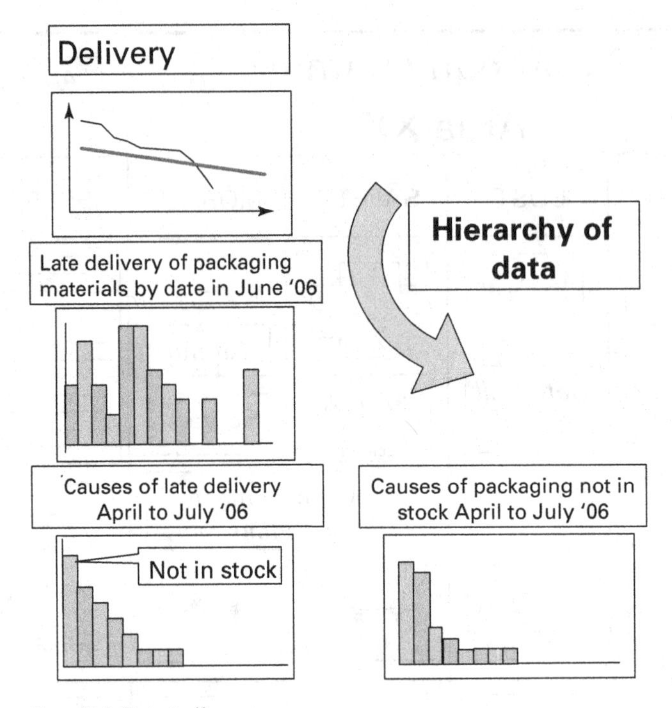

Figure 5.4 Data trail

of cost performance, as it tracks the losses of the production area which can easily be turned into real opportunity costs to the team with sub-optimal performance.

The communications board (figure 5.3) does not just impart the graphs which show historic performance – that would only tell part of the story needed for the team members. In addition to each measure is typically found a 'cause and effect' chart for safety, morale, quality, delivery and cost indicators and a 'data trail' for current problem-solving activities. These added pictures tell the observer where the main problems can be found with each generic measure and serve to focus problem-solving initiatives by identifying the main 'culprits' to investigate. So you may find that one of the branches on the delivery 'cause and effect' chart is the late delivery of the production schedule to the area, or the lack of packaging materials available (Rich, 1999). The data trail then uses graphs and numerical data to help analyse the causes. Figure 5.4 shows the detail of the data trail from the communications board shown in figure 5.3.

The team wishes to improve delivery and from their cause and effect diagram they know late delivery of packaging materials is an issue. To try to understand the issue, they plot when late delivery occurs and also gather data on the causes of late delivery and plot it in a Pareto chart. There appears to be no cyclical data on the graph that shows data by date by the largest cause of last delivery is 'material not in stock' so a further Pareto graph is plotted that shows causes of materials not in stock. In this case this is the end of the data trail and the team can take action, but for other problems the data trail may have several layers.

Beneath these charts, which now show the performance of the team and the results of team brainstorming and problem solving, other charts which show a list of what projects have been launched to solve and eliminate the causes of problems. These items show what the project is, the team members involved and the current status of the project (launching, on-going or completed). Now in one vertical sweep of your eyes you can see the performance of the team, the likely problems, if achievement is less than 100% and what projects have been developed to keep the improvement momentum going. So, in the case of a poor delivery improvement focus, the team may elect to introduce a pull system for packaging materials or locate a printer at the point of production so the schedule can be printed directly in the team area rather than being delayed by the internal mail. Tracking your eyes horizontally therefore lets you know how the team is performing on its other indicators and what projects are active. These measurement systems also form a 'scrap book' of past projects so that even the hardiest of cynics can see that change is happening in the production area and by their co-workers acting together.

Another benefit of the communications board approach is that the key measures – the headline items of safety, morale, quality, delivery and cost – will not change over time and are in fact the key measures that are displayed at internal customer production areas and internal supplier areas. So now you can tour the factory and see information displayed in the same way – sure some of the actual measures may differ, but generically the system operates in a common manner and wherever you stop you can take a quick scan of the board, and health check that team's performance. In addition, when team briefs are conducted or visitors are shown around the factory, these communication boards become the venue for the brief, so people can gather around and see the facts about what is going on in their production area.

At one Learn 2 company ('Cosmetic') these communication boards were coined 'dashboards' because of their role in planning, controlling and navigating the improvement process by a line operator and this term stuck. It was, given what has been outlined in this chapter, quite a good analogy – you cannot drive a car effectively without periodically checking the control instruments on the dashboard. This one point of information shows how many miles have been travelled, whether the engine is getting abnormally hot, whether a warning light has come on, the status of the fuel, whether you are speeding or whether you have left the indicators on. The analogy is a good one – one central point of information, which, if you are driving, gives you a huge amount of data from just one quick periodic glance. Regardless of what the mechanism for making transparent the performance and trend information of the team is called, the most important lesson here is that the 'visualisation' process, which began with workplace organisation, is continued and includes the 'visualisation' of performance feedback.

Box 5.1 Too many measures! What's the most important?

At one of the Learn 2 sponsor companies, it was found that the team collected and reported to management 23 measures of performance on a regular basis. Collecting and processing this information was hard and time consuming and most of the measures were not those associated with the routine management of the production area. In addition, no-one, not even the managers, could identify either the most important measure or the key projects that had been launched to improve each of the measures that were collected – worse still the most important measure changed each time the team met with their area manager. One quick session and many of the measures were removed as unnecessary or not used by anyone. The remaining measures were then assigned to the key measures of safety, morale, quality, delivery and cost performance to result in a simple-to-understand system of visual reporting, easy data collection routines and a logic to the workplace which did not exist before as every measure seemed as important as the next and no-one knew which to concentrate upon. The communications board was implemented within a single day (borrowing a disused notice board) and remains there today.

Outside the canteen of a Japanese-owned manufacturing facility in the UK is a set of a dozen measures that show how well the factory is performing relative to the plans established by the management. Each measure has a line chart which shows performance, but workers would need to be within a metre of the board to read each chart. To assist with communicating the status of each measure, the management introduced a series of simple 'smiley faces'. These faces were printed on to circular shapes of yellow coloured paper about six inches in diameter. A smiling face was attached to any chart where the performance of the factory was better than the expected standard, a frowning face when the measure was satisfied and a sad face when the performance of the factory was below the standard needed. It was a simple and effective way of letting every employee know how well the business was doing.

Visual management was practised extensively at the Learn 2 companies involved with the research programme. It was not a simple lesson to learn or teach. In many respects business managers perceived visual management as somehow less exciting and interesting than other elements of lean production with which they were familiar but had not implemented (like kanban systems). However, following the success of the CANDO and 5S change programmes, VM was regarded as a simple extension of the workplace organisation and accepted by line workers as the next stage in learning how to implement lean production.

Box 5.2 The daily walk

At a Birmingham-based automotive manufacturer, an entire room is dedicated to a performance storyboard and there are many team boards around the room which start with the customer and follow through the finishing processes to the primary conversion activities of the firm until the last board reports the performance of the supply team. At the beginning of each shift, the team leaders report the progress of the factory and any deviations which have been detected. Every day, all managers attend a morning meeting, lasting about 30 minutes, at which the Managing Director is 'walked through' the entire value stream of the factory. Interestingly this meeting room has no chairs (not even for the MD), so that everyone involved with the meeting has to move with the discussion.

For managers the communication boards also hold high levels of value. Managers do not think in the same timeframe as teams. Teams think about the day and the week, whereas managers tend to think about this week and the calendar month. Senior managers think about quarters of a year and several years ahead. This differentiation results from planning and the need to review meaningful information. If a senior manager was to review performance on a daily basis, then the information would be confusing – they need summary and trend information because their job is to guide the direction of the business and to identify the weak areas within the entire factory so that additional effort/resources can be seconded to the team concerned. So the communication boards serve another very important purpose, they are at the bottom of a very important feedback system for senior management and at one of the Learn 2 companies' managers chose to have their own communications board where the data related to their higher levels of information. In this manner the entire value stream performance can be averaged by combining all the different production area boards and reporting this information. For managers, trends can be established which help to determine what can be offered to customers and how the team improvements can be harnessed to prevent competitors from attacking the customers of the firm and also as a means of finding out how to increase business by targeting new customers. So the communications boards help senior managers to assess 'where the company is' as well as targeting 'where the company needs to be' by predicting what the market will need in three years time. When senior managers plan the business by looking at what the market and customers will expect in three years, then you are likely to find changes needed in the quality, delivery and cost performance of the factory, and, as it is better to take action now than to wait until these market pressures force a reaction from the firm, this information can be used to identify targets for achievement by each and all teams in the factory.

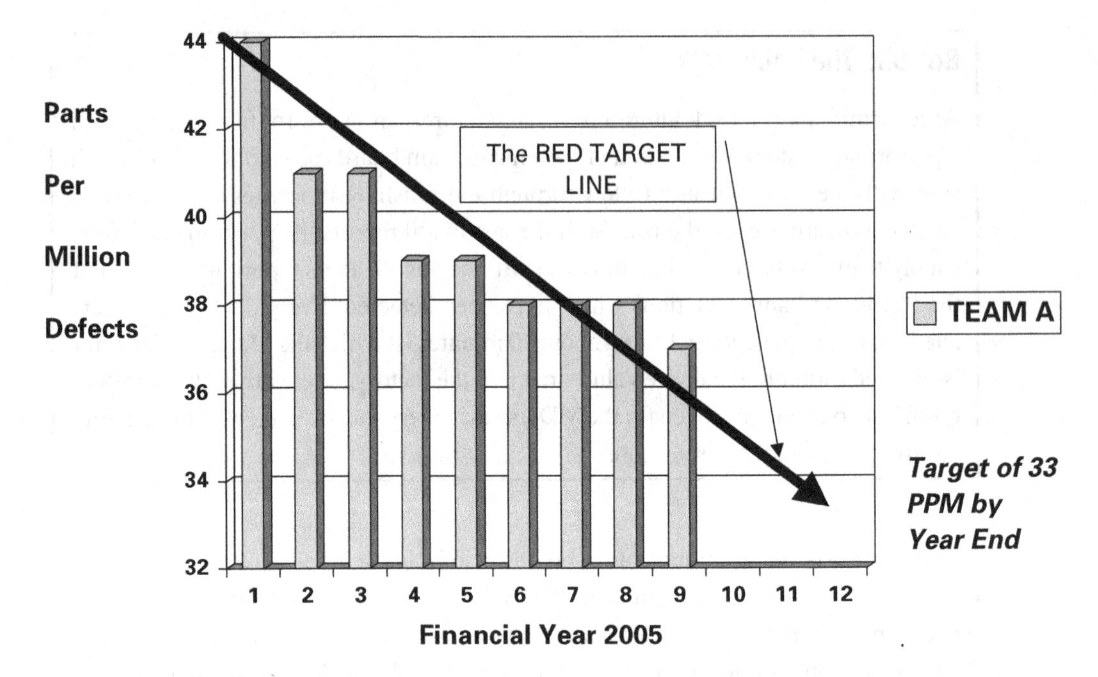

Figure 5.5 Graph with target

Most advanced lean businesses therefore add a target achievement line to each key indicator measured by the communications boards (figure 5.5) and this adds a new power to the team – it shows drift between team performance and the annual goal set by the business.

The new charts therefore link the team to the factory and the team with future customer expectations, so improvement activities conducted by the team can be related to 'winning new business' in a way that never existed before. It also serves to highlight, through the performance graphs collected by the team, which is the weakest area of their manufacturing self-management and where to invest in improvement activities for the most gain.

The process of linking future-directed improvement targets and goals with the feedback routines of the teams is commonly known as policy deployment and this will be covered in a later chapter. Policy deployment is one of lean manufacturing's best kept secrets, as this process of target setting and review is a hidden element of how the total production system is managed and directed

Summary

Work standardisation and a disciplined approach to workplace organisation are key to equipping employees with the basic improvement skills that can be nurtured for later stages of development in terms of improving quality and delivery performance.

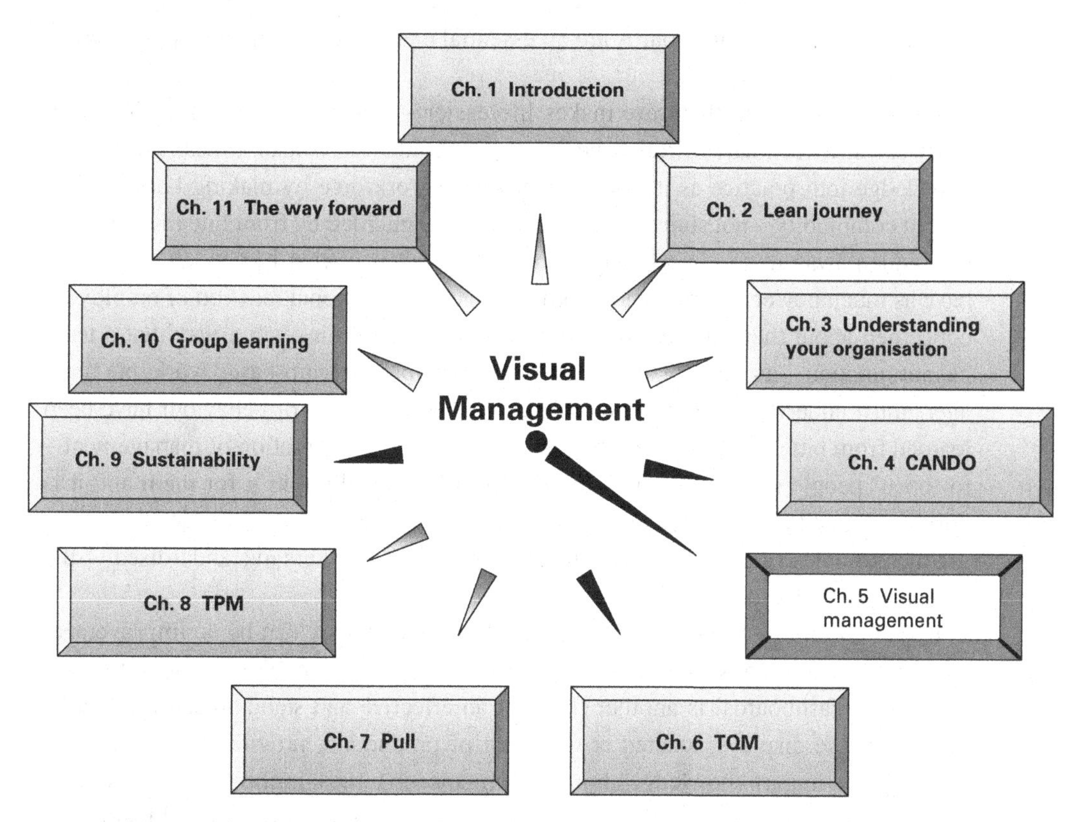

Figure 5.6 Visual management linkages

However, very few lean activities really ignite and integrate the workplace teams more than getting people involved with visual management – it is a new mindset and way of thinking that is easily transferable to all manner of project work and truly embeds a spirit of simplification and standardisation in all factory projects. However, visual management is not simple; it is an art, and a very good science. Visual management is best experienced. The real trick is to treat every case of visual management application as a challenge for simplification. Once again the process that will lead you to your eventual solution is the practising of problem-solving skills in the same way that these skills were used for CANDO but were not necessarily taught during these phases of lean development. Most team members get a real 'buzz' from visual management and most factory auditors/safety representatives and trade unions like to get involved too. From just a single activity, the lessons can be scaled up to the entire factory level. This should include the setting of factory-wide standards that create a common approach (e.g. a common colour scheme) so workers can move between areas and still understand the basic system of work flow and control (e.g. green-taped floor space to hold finished goods, yellow for work in process, blue for raw materials, orange for change parts).

In addition to CANDO, visual management has links to many other chapters of this book (figure 5.6) including TQM, in terms of highlighting the normal and abnormal, and effective communications boards are an essential element of sustainability – a subject dealt with in a later chapter.

Visual management therefore makes life easier – it is an investment in time with many rewards for many different business stakeholders. It is also a means of getting knowledge into practice as it brings together the workforce by making knowledge a shared commodity – not something that is jealously guarded by front line management nor hidden away under the assumption that, 'If I tell you what I know, then you will have as much power as I do.' Hiding knowledge is a traditional reaction of employees concerned about their job security, protecting information in such a way that a team cannot function – its actually against the very principles of team-based work, but managers must understand this form of behaviour. These patterns of behaviour have been learned from bitter past experiences and often a previous generation of management – to 'open' people's minds to sharing there must be something in it for them and it is not that difficult to find ways of moving people with the skills to a new level (by giving them more skills), whilst letting them train their co-workers in standardised visual management techniques.

It is often said by businesses on the lean journey that there can be no improvement without these types of standards and the visual communication of information through the use of 'dashboards' is another 'hub' of an effective and standardised production system. These displays, located at the point of production, provide information not opinions and impart knowledge about performance in a meaningful form for the team and visitors to the area to easily digest. The communications board should show the performance of the area in terms of a few key measures and additional team and safety information. The board should not look like a 'sweet shop' window with many different measures – it should focus improvement on the vital few.

6 Problem solving, TQM and Six Sigma

Ann Esain

If you work in industry or commerce today, then the topic of 'quality' will be part of your business language. Quality is an order qualifier, something demanded and expected by customers and the cost of poor quality is quite often a measurement at the highest level of an organisation (Hill, 1985). This is not surprising when figures such as 40% of the total cost of doing business is attributed to avoidable activity due to rework, scrap and rejects. Known as the 'hidden factory', this describes the multitude of, mostly, small activities which make up a significant cost to business, but are unnecessary. A waste of resource, worse still, a source of frustration between employees and across the different levels of an organisation.

This chapter will briefly outline the role of quality in a lean transition and its place in supporting such a transition. We are aware that the concept of quality is discussed in many guises and often as 'initiatives' by management authors, and, for the most part, these initiatives are considered old and staid. Quotes, from line workers and managers, that 'we did TQM in 1996' are not uncommon and reinforce the idea that this type of approach is a 'tick in the box' exercise. However, 'world class' businesses, and those determined to become world class, know well that quality management lies at the heart of all improvement activity.

We know that to maintain performance we need to continue focusing on a task. This includes attention to quality, just to stay at a consistent level of performance, let alone improve further. Quality may be an old idea, but sometimes these are the best and in our view attention to quality is integral to achieve the performance levels associated with lean.

This chapter will deal with the question, 'why quality?' The human capacity to solve problems is a huge asset and will be discussed further. How an organisation can achieve high quality and sustain high performance improvement programmes will also be addressed. As with previous chapters, we will not discuss in detail all the tools which an organisation will need to undertake change, more the sequence, capabilities, with tips and hints to help the reader think about the contingent needs of their own organisation. Finally the chapter will introduce Six Sigma as an approach to quality and problem solving. As a trained Six Sigma black belt, with first-hand experience of Six Sigma and of lean, the debate of '*Lean* v *Six Sigma*' has been interesting to observe. Indeed, such a debate misses the point: it is not what the tools are called, but the individual problems of the organisation which need to be addressed. The real message from our work is that Six Sigma and lean are compatible, assume integration and come from very similar roots.

A valuable learning point is that it is best to start with a simple approach, identify the quality issues that need to be addressed through mapping (see chapter 3) or some other analysis tool and then provide time and training to enable the workforce to resolve the issue(s). This approach enables a cross-functional group to agree on the priority of the issues revealed in the organisation or supply chain. These issues can be dealt with in order and resources can more easily be identified.

Six Sigma relies on statistics to resolve problems. So, using the full range of the Six Sigma tool set should be reserved for the complex and strategic (George, 2002). Do not use a hammer to crack a nut. The skills associated with Six Sigma and the statistics employed are a very valuable asset to your organisation and can help blow away myths about many issues through the analysis of data. Like all tools sets, they should be practiced in a safe environment that enables the individual to become competent. When individuals become experienced in a variety of tools, they will select the most appropriate to solve their specific problems and issues. A word of warning, do not train for training's sake; train when the time is right and link it to the issues to be resolved. Six Sigma training is not a numbers game – having more or less will depend on the strategic needs of the organisation. So this assumes that there is a mechanism in your organisation for the selection of strategic and tactical learning projects.

How you can do this will vary based on the size of organisation, skill levels etc. Six Sigma focuses on cost as its key performance measure. This helps to produce cost–benefit analysis of the up-front investment in training (return on investment, ROI). However, the measure of return on capital employed (ROCE) should also be used as a measurement. Quite often organisations lose the benefits of freed-up resources because there is no strategic approach to redeployment. Hence, there is individual project evidence of ROI, but this is not reflected in ROCE. This is key for all change programmes. Efficiency savings are often not useful if they are not matched with an increase in sales, e.g. freeing up space will not help ROCE unless it can be released for an alternative profitable use.

Quality and Six Sigma in practice

To really know if quality is an issue for your organisation you need to know what the current state of your processes look like. As mentioned in chapter 3, value stream mapping provides a diagnosis of the current state of your organisation and processes (Hines and Rich, 1997). It helps determine the direction of change through the engagement of the shop floor, and focuses attention at a management level on the vital few, by visualising the biggest issues and indicating in which order these need to be remedied to create a 'best in class' business (see Box 6.1).

Box 6.1 Air Repair mapping

Perception is as powerful as facts in some organisations – but opinions are often misleading and dangerous. Initially, a tool known as VALSAT was used to establish what wastes were most prevalent in the processes (Hines and Rich, 1997). Interviews were conducted by the improvement team with different individuals drawn from different departments connected with the process. In this case the waste of defects was clearly considered to be the most significant (see below). The perception was clearly that quality was a key issue for this organisation.

WASTE	John	Kevin	Nick	Peter	Phil	Colin	AVERAGE
1 Overproduction	3	0	3	4	0	5	2.50
2 Waiting	6	10	4	0	10	5	5.83
3 Transportation	3	0	5	1	0	8	2.83
4 Inappropriate Processing	2	5	5	7	10	0	4.83
5 Unnecessary Inventory	6	5	3	3	0	2	3.17
6 Unnecessary Motion	5	5	5	10	5	6	6.00
7 Defects	10	10	10	10	10	9	9.83
Totals	35	35	35	35	35	35	35.00

An exercise was then undertaken using a quality filter map and as figure 6.1 shows the issue of quality was a clear problem. Indeed the team identified that poor internal quality was the root cause of poor delivery. Internal quality issues were related not only to the parts but also to the paperwork being used.

Hence the approach for this organisation was to combine CANDO (5S) with problem solving to move towards a vision of being more competitive in the market, which in turn would provide security of jobs. The team identified a desire amongst the workforce to be recognised and valued for their contributions; there was a need for a continuing improvement programme and that this programme should be supported by the senior and middle managers of the firm. Further, the exercise found a general interest of every employee in understanding what progress was being made against targets and to make these targets visible in the factory for all to see.

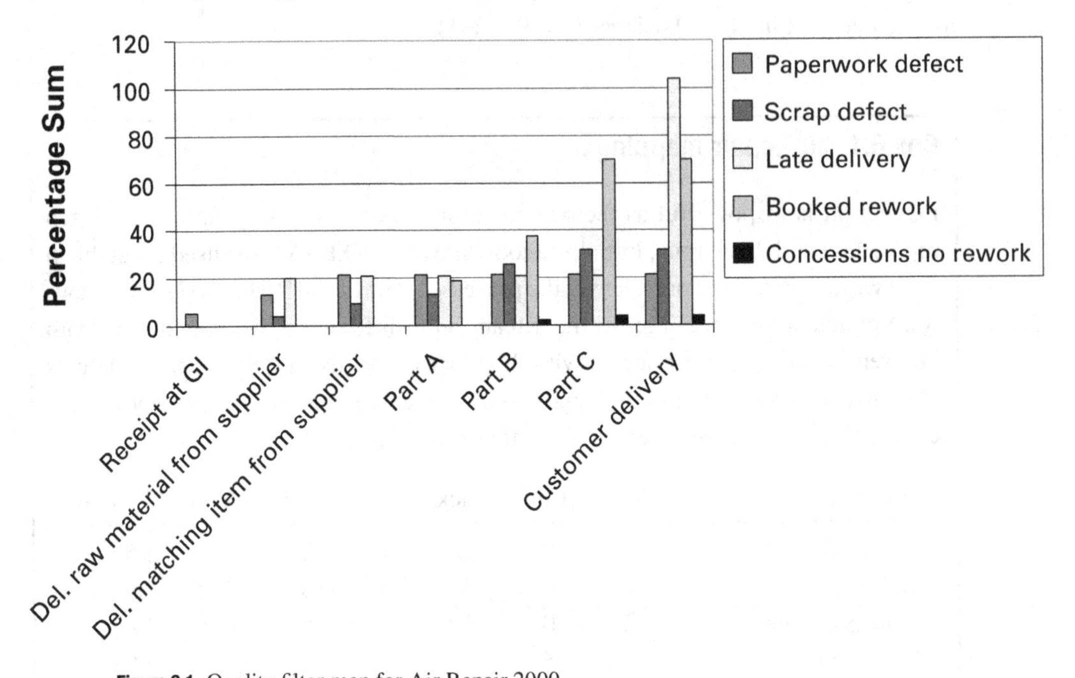

Figure 6.1 Quality filter map for Air Repair 2000

Problem solving

The human race is blessed with the capability to solve problems. If there are no life-threatening problems to resolve, we devise games and pursuits which oil the mechanics of our brains and give us the pleasure of being able to resolve the issues presented. We only have to look at the numbers of people who do crossword puzzles, as an example, to understand this capability.

The capability of problem solving is an inherent ingredient in all change programmes and increasing and refining that capability is an important step. By undertaking regular problem-solving activities these skills become more honed. They become an unconscious asset to an organisation, what some refer to as an intangible asset. A lean implementation programme will raise many problems. Hence purposefully capturing and enhancing your workforces' ability to solve problems is key to enabling a lean transition.

Problem-solving tools are designed to identify root causes of problems so that a problem cannot reoccur in the future. This needs time. Many organisations fire-fight to maintain the status quo and consequently display a habit of applying a sticky plaster because, on the face of it, it seems the fastest way of resolving the problem. Usually

Know Your Organisation & Your Customers

What		How
Leadership Reason & need *Why should this organisation undertake change – What is the problem? What are realistic expectations? What measurements need to change?*	Executive Unlearn to Relearn Forward 3 years	Leadership Belief & demonstration *Consistency of message & walking the talk* *Expectation depend on resource availability, pace people can change, product cycle time, etc*
Leadership Embeddedness *Provide framework of logic for how to resolve the problem (s)*	Management Unlearn to Relearn Knowing to Doing Forward 1 month to 1 year	Leadership Logic *Resource – to demonstrate* commitment *Co-ordinate and control – for ownership & measurement*
Control participation *To agree problem and assess different solutions to resolve the problem* *Implement, learn and adjust*	Supervisors and Operators Unlearn to Relearn Knowing to Doing Daily – Weekly – Monthly	Measure and adjust *By creating feedback loops against action/ implementation*
Where are we now?	Value stream mapping	Where do we want to be?

Figure 6.2 Problem solving at the different levels of an organisation

the plaster falls off and the problem costs more in the long run and creates a morale issue between staff and managers (see figure 6.2).

Problem solving, in the business environment, is often difficult, not because the human is not capable, but because the first step of defining the problem can become a struggle. Have you ever been sent away on a training course and on your return been presented with the most elusive problem the organisation has to solve, because somehow that training equipped you with special powers no one has had previously? This is certainly an experience of some individuals who have undertaken TQM and Six Sigma training.

The first step in problem solving is defining the problem. Figure 6.2 shows how problem solving is integral at all levels of the organisation. It also illustrates that the timeframe for problem resolution is different at different levels of the organisation. A key question for the organisation is, 'At what speed do I want this change to take place?' Resources then need to be made available to meet that speed of change. If the resources cannot be made available, then the speed of change needs to be adjusted.

One of the advantages of Six Sigma is the way in which problem definition has its own set of tools and is classified as an element of the approach in its own right. The step of 'plan' in the plan, do, check, act cycle (PDCA), popularised by Deming (see chapter 9 for a full explanation), relates more closely to the stages of measure and analyse in Six Sigma. The define stage in Six Sigma requires not only the definition of the problem but the creation of a charter which is linked to the strategic direction of the organisation through some sort of approval mechanism. This stage requires the identification of roles and responsibilities and the anticipated result in monetary terms.

To define a problem the individual or team usually needs data. Our experience is that at the outset these data may not be available in the format required and will necessitate the use of perceptional data. Tools such as FMEA (failure modes and effects analysis), the cause and effect prioritisation matrix and quality functional deployment are designed to be able to accommodate this type of data. Indeed the principles of cause and effect are very important, as often it is the effect which becomes the focus of the problem solving exercise rather than the cause. This can result in creating inefficient solutions. Similarly, while the human being has the capability for problem solving, it does not mean that we select the most effective solution to resolve the problems we are presented with. More often than not we select the most elegant given our knowledge and do not 'sweat or torture the data' to see the problem from all directions, learn and reach out to deal with the root cause. It is a balancing act because we also observe that the data analysis stage can go on forever (paralysis by analysis: this is considered to be an intellectual phenomenon – the more highly qualified, the more analysis and discussion before implementation) whereas 'having a go' would give much clearer understanding.

Box 6.2 Medical Devices case

Following intensive training on both lean and Six Sigma, the groups were set a final exercise to combine all the knowledge learnt to establish how a manufacturing plant producing products for the domestic market could achieve 100% quality, 100% delivery at the lowest cost. This problem had many facets and one of the early tasks was the identification of the problems in the production process. These production issues were exacerbated by the product having a number of features which were 'critical to quality' performance and if not strictly managed caused the product not to function effectively.

The group, having been given this task, dutifully reorganised the flow of manufacture of the product, but did not produce a product which was of an acceptable quality standard. Hence all the products made were rejected by the customer. This is a typical reaction to this type of dilemma. Reorganisation seems the most logical starting point. In fact the investment for reorganisation has been made without, at this point, any return, because the quality issue was not identified initially.

> Realising there was a quality problem, the teams drew upon their training. 'Let's design an experiment', we heard. The fact that, having played the game for several rounds, a lot of very useful information, which had already been collected and available, was missed. The idea of keeping things simple is really important as is starting with what the customer wants and values. Both points are often missed in the exercise and in a real setting.

The approach of CANDO relies on problem solving and seeks to sharpen these skills in the workforce (see chapter 4). Problem solving requires tenacity from the individual and the team, alongside creativity.

Good problem definition is key and as more data become available, do not be afraid to amend the questions posed, but do be aware of '*scope creep*'. Scope creep is where the problem being addressed grows until it is too big. Always try to keep the problem small and manageable through the organisational support structure. It is better to have a large number of problem-solving activities than too large a project. Measurement is integral to good problem solving. Otherwise, how do you know if what you have done has made a difference?

If you are new to change, focus on something you will have some success with. Success is the calling card, the proof that teams can deliver against the harder issues in the organisation and supply chain. Being new to change, we often miss that we also need to learn how to use the various tools available and this is why you should select a problem or area which is small enough to ensure success and learning. Once you are familiar with the tool sets, move to those things which matter to the organisation and supply chain. Link the problems to the customers' wants and needs, the strategy of the organisation and the 'hot spots' which may well have been revealed by mapping (chapter 3). If problems are random and resolution is not linked in this way, conflict can arise when it comes to asking for time and effort from other areas. We often see difficulty in gaining support across functional groups because of differing priorities, and the only mechanism for resolution is the senior executive team (see figure 6.2). If the problem is not aligned to what the organisation is trying to achieve, the resolution process will fail, leading to frustration. The need to align projects is highlighted by Motorola in the book *The New Six Sigma*, where they have introduced the concept of a balanced scorecard to align their projects to the business objectives in a highly visible, published and easy-to-understand way. They have introduced four stages, 'Align, Mobilise, Accelerate and Govern', as their new Six Sigma route.

Why did we start with problem solving? That is easy. To achieve quality processes and products, the practical skill of problem solving is paramount. This capability is available to your organisation through every member of staff. We just do not use it. If we use it properly, make it a standard approach, then our ability to succeed is unlimited.

Visualising the problems and displaying the statistics are a great way to help the workforce see progress – fast and slow. It is said that about 15% of all problems in organisations can be solved on the shop floor alone. Variation, which is present in all processes, can be divided into common and special causes. Deming suggested that the special cause variation could be solved by those in the workplace, whereas common cause (natural variation) requires management intervention. Hence a meeting on the shop floor at regular intervals, usually daily, with those functional groups which support the shop floor in delivering product is often adopted. The assumption here is that the shop floor is the place where the pace of customer demand is satisfied. Therefore the problems at this point are in need of the most urgent resolution. Another suggestion is that managers should make it their task to stand in an allotted space and watch what is happening. This approach is designed to stop the frenetic activity of managers and build an understanding of the detail behind the problems, which occur on the shop floor on an ongoing basis.

Organising a regular time for dealing with the daily problems recorded by the supervisor and his team, as well as making this visual on the shop floor, sends out a powerful message that the organisation is serious in jointly resolving problems with the workforce. Quality guru's have clearly indicated that the presence of management at the place of production is critically important to improvement success, Peters and Waterman (1982) talk of management by walking about, for instance. This means that a number of problems can suddenly emerge (hidden because operators had given up trying to share their concerns in an organisation where a process for resolution is not formalised). Volume of problems can become an issue which, if not managed correctly, will be counterproductive to change (just like not dealing with Red Tags in CANDO, see chapter 4). A way to deal with this is to empower the supervisor and his team to prioritise problems which need to be addressed first. This can be undertaken drawing on the previously used cause and effect prioritisation matrix and possibly a failure modes effect analysis (FMEA) alongside a fishbone diagram (see figure 6.7) and pareto analysis. All can be displayed to communicate what is happening.

In addition, a problem resolution board summary can be used (see figure 6.3) to share progress, both in terms of resolution and in terms of speed of change. This is important because of the understandable cynicism of the workforce, which is usually based on years of experience where nothing gets changed. If progress is seen, those less enthusiastic may well start to feel that the organisation is listening (the start of communication). This is also a great management tool. As a manager, you will not have to ask questions about what is the current state of the production line, as this is visual. Instead the time can be spent understanding what needs to be done to resolve the issues highlighted by the boards. These should start to identify the root causes of problems in your organisation and supply chain. Like all visual tools, only use them if you are prepared to consistently update them. To start and then not continue leaves a message that the organisation is not serious about improvement.

An example of problem-solving methodologies is that used by Ford and an approach called 8D meetings. This enables the specification of an immediate action (the 'sticking

	Problems Today	Awaiting Action	Closed
No. of problems this week			
Cumulative no. of problems			
No. of reoccurring problems			
Review the rate of improvement			

Figure 6.3 Problem resolution summary board

plaster') and then the long-term resolution, so that the effect never occurs again. These meetings are not simple, but enable a robust system to be established. They require discipline (adherence to standards) at all levels of the organisation, as well as from functional groups.

Quality matters

Without quality processes, organisations and supply chains cannot achieve optimised performance. Quality matters because poor quality can lead to the failure of the implementation of lean principles and techniques. It is assumed by most texts that good quality is in place before a lean transformation is attempted by an organisation. As a generalised rule an organisation with a product/service failure rate greater than 10% should think about a quality initiative as an early step alongside workplace discipline CANDO (which should have helped to identify and solve some of the route causes of the failures that occur).

The aim of the lean approach is to create processes which will indicate if an abnormality exists in that process (see chapter 3). A quality issue is an abnormality. Whilst there is much discussion about the improvement initiatives around quality and there relative merits, without a positive approach to quality (regardless of what it is called) a transformation will be limited. When do you know this transition has occurred? It is the point where problems and issues are welcomed and openly discussed by management and workers alike.

There is a widely held belief, demonstrated by the language and actions of the organisation, that problems are 'somebody's' fault. It is hardly ever that simple. Deming and Juran claimed that management are responsible for the majority of quality problems

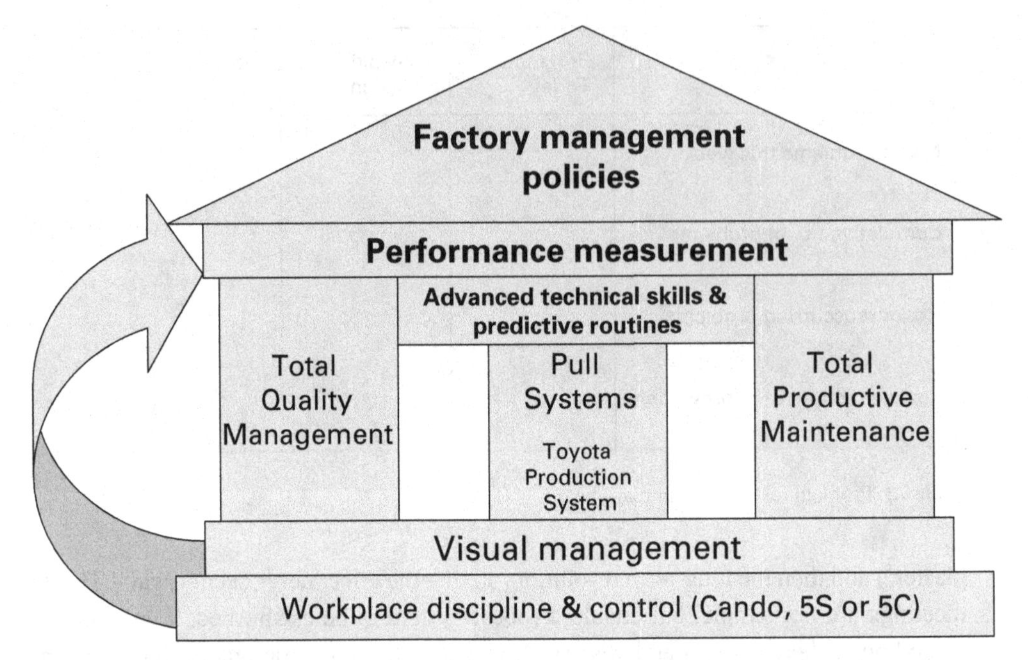

Figure 6.4 Problem solving is integral to the achievement of improvement and change

(94% and 80% respectively), however this language is equally unhelpful (Bicheno and Catherwood, 2004). If problems are due to a common cause variation, then the problems which need to be dealt with will require cross-function participation and resolution. Existing methods of target setting, objective setting and management styles (an example of which is 'bring me the solution never a problem') have a lot to answer for. The fire-fighting seen in organisations is usually as a result of inappropriate behaviour and rewards, which reinforces bad behaviour at all levels of the organisation (playing Deming's (1986) red bead game will demonstrate this).

What is quality?

The idea of quality has been in existence for many centuries, but the importance of quality has come to the forefront of management text since the industrial revolution, driven by the need for interchangeable parts. The premise was that skilled craftsmen would be able to predict a need for repair if the parts were interchangeable. Each worker makes one specific part, or performs a limited number of operations on a given part to exact specifications, so that all the individual parts would be identical and could be assembled into a product. If the parts were interchangeable, then good quality had been achieved.

So quality is the ability to repeat actions, which means that a part, product or service is identical every time (Rath and Strong, 2004). This can be likened to a certain company producing burgers and wherever in the world the specification is identical. The burger is standardised. There is a difference in having standards, being able to interpret these standards, being able to keep to those standards and being able to measure those standards. This may explain why so many people see quality as different things. Sony has developed an approach called windows analysis which tests if the appropriate standard has been developed, communicated and adhered to.

Whilst quality is explained here in the terms of how an organisation achieves it, there is an additional and more important aspect of quality, that of the customers' view. In the five lean thinking principles, this is the 'value' stage (Womack and Jones, 1996); that is, what the customer values in the product or services which are being supplied. In Six Sigma the voice of the customer is the key enabler to identify what their wants are. This is then used as a specification, which is known as critical to quality.

The word 'quality' itself has different meanings for different people, dependant on their experience in the area, their sector, there progress in dealing with quality issues, etc. The following different levels of quality represent a path for the evolution of quality within an organisation.

- Level 1: Inspection – which can be the entry level to manage quality issues
- Level 2: Quality assurance – moving from detection to prevention
- Level 3: Total quality control – with the control system in place, start to unravel root causes of quality problems, linking them to internal and external customers
- Level 4: Total quality management – which takes quality to the heart of the business and brings quality to the top table as a management issue. At this level the cost of quality is measured.
- Level 5: Six Sigma – currently very popular and is being heralded as the improvement methodology for all organisations. This approach, developed in Motorola, has a comprehensive structure in place to deliver bottom line cost improvements.

Note: Level 3 – Total quality control would be considered the entry level for a lean transition.

The quality movement in the early twentieth century provides the basis for many improvement philosophies; this is the case for all the levels of quality discussed above as well as for lean thinking. Figure 6.5 shows the interlinking nature of quality with other sections of the book

Inspection

There is a belief that, if an individual inspects or checks someone else's work, then errors will be identified and the cause will somehow miraculously disappear. Have you ever heard the comment, 'Why hasn't someone done something about this?' In

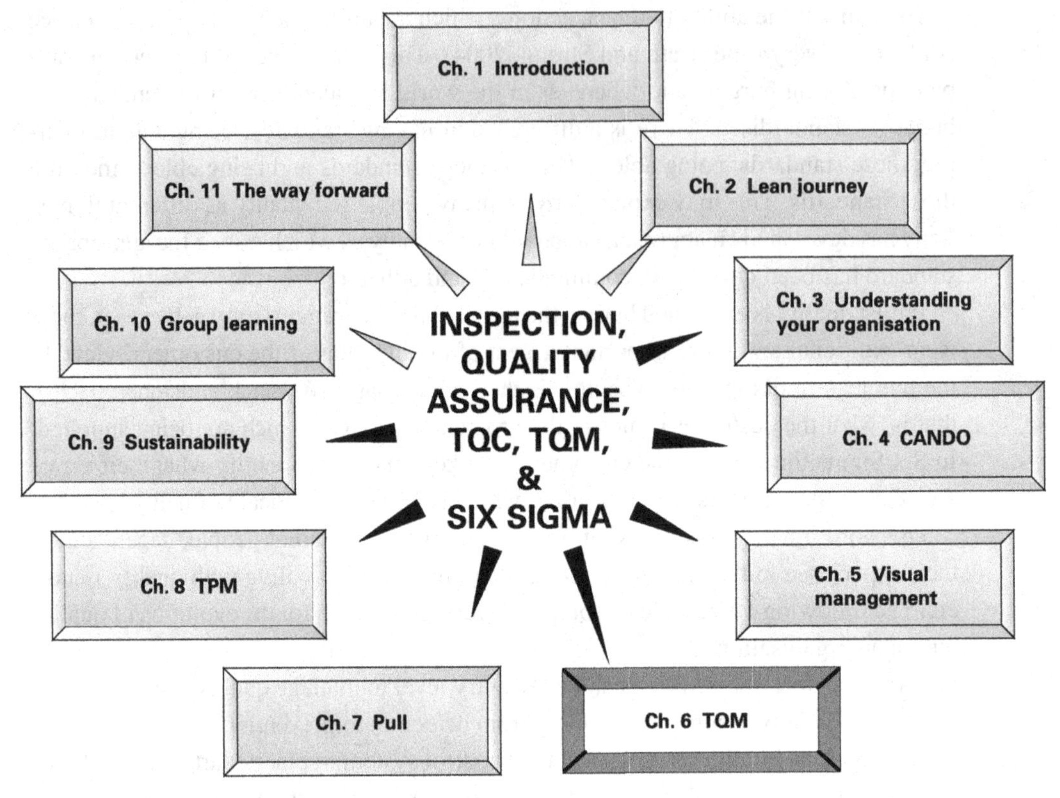

Figure 6.5 Quality links in the lean journey

truth, at this stage of quality control all that normally happens is that the product is sorted and salvaged. There is unlikely to be any feedback to the sources of the error or the omission. Whilst this is an important mechanism for collecting data about the magnitude of something which is going wrong, it does not provide a system to enable change to be enacted.

Inspection should mean that there are standards against which to inspect. Standards at this point are often based on samples only. However, in this early stage of development, it may be that the standards are not sufficiently robust to ensure a product or service is delivered identically each time. The person(s) used to inspect the product or service should therefore be more skilled. This then introduces a power dynamic, which introduces a dilemma for the inspector. To make 'simple and clear' those things which need to done, could ultimately mean that the inspection job could be at risk . . . and no one is likely to improve their job to the point it is no longer needed.

Hence the inspection level of quality is useful to understand more about quality problems, but is unlikely to improve the organisation's bottom line performance. In public sector organisations, this reliance on 'audit' only identifies what is incorrect not what can be done to change the situation. This in turn exposes organisations and leads to the suppression of information rather than the use of the organisational energy to make improvements. The description of activities in level 1 reflect the mass paradigm of production, where tasks are cascaded down the organisation and improvement is a separate activity to the tasks individuals complete. It is clear that to become a quality organisation steps need to be made to move beyond inspection.

Quality assurance

Quality assurance is the first real shift from detection to prevention of quality and the idea of built-in quality. Inherently, if someone checks your work, it is unlikely that you will check your own. Indeed feelings espoused by those whose work is inspected by others include lack of trust to do a good job, therefore, at its extreme, 'why bother with doing a good job'. Hence in quality assurance, inspection activity is returned to those who are producing the product or service, whilst the control aspect is monitoring the effects of poor quality. It also returns the responsibility for both producing, maintaining and improving the product to those who make the item. This usually requires an improved production standard to inspect against. The control aspect enables a common set of standards to be developed, and maintained, but training in these techniques is against the standard.

Also the introduction of pre-emptive activities, such as calibration of equipment, will become part of the quality department's activities. Whilst important, these improvements are based on following a set of rules. These rules are often set by manufacturers, rather than for the application for which the equipment is being used. If there is an understanding of measurement systems analysis, then refinement in both standards and equipment management can be more effectively achieved. Measurement systems analysis will be discussed later in this chapter.

A mistake common at this stage is dealing with the effect not the cause of the quality problem. The seven basic tools of quality (Bicheno and Catherwood, 2004) can be used to help detect the quality problems in the production of a good or service. Quality assurance uses the quality system and the seven basic tools of quality as a natural part of activity within the organisation. As a result the quality issues which remain are often the more complex. Complex issues are those with more than one variable which will relate to the root cause of a problem.

In an organisation or supply chain where quality assurance is practiced, quality levels of performance are known and understood. The cost of quality both in terms of detection and prevention are understood and activity is focused on insuring that quality issues are detected at source and never get passed on to the customer or consumer.

The intention is that a wide proportion of employees start to understand and apply the seven basic tools of quality (see figure 6.6). Indeed Dorien Shainin in his book *World Class Quality* refers to the basic quality tools as kindergarten tools. Shainin's assertion is that only by applying tools which give us a completely new insight can process knowledge be used to transform processes. These tools enable clear categorisation and

Process Chart

Pareto Analysis

Ishikawa Diagram

Histogram and Measles chart

Run Diagram and Multi-Vari Charts
(this includes Statistical Process Control charts – SPC)

Correlation and Stratification

Check Sheets and Tally Charts
(this includes Standard Operating Procedures)

Figure 6.6 The seven basic tools of quality (Bicheno and Catherwood, 2004)

start to challenge the status quo. Employees need to know how these tools can be used to establish root causes of problems (see figure 6.7). The outcome is that time becomes available as a result of removing steps previously used, i.e. free up capacity. This then leads to further change to deliver improvement and bottom line cost reduction.

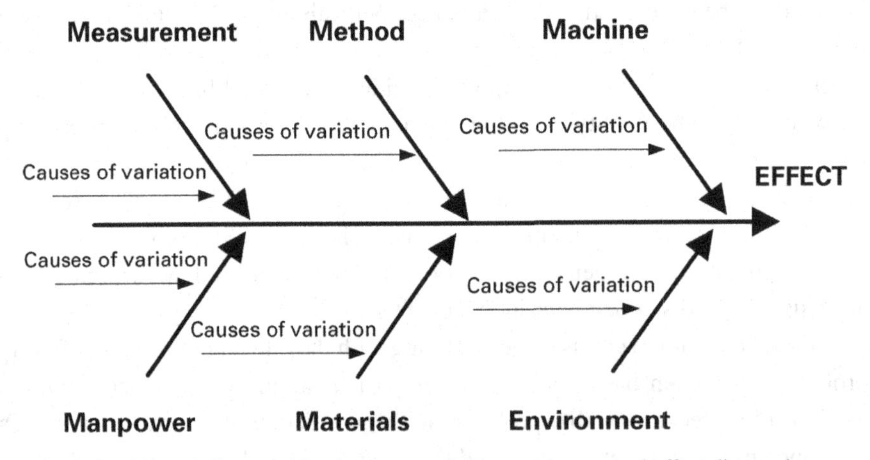

Figure 6.7 Ishikawa diagram (also known as a fishbone or cause and effect diagram)

Total quality control

Feigenbaum (1991) defines total quality control as 'an effective system for integrating the quality development, quality maintenance and quality improvement efforts of the various groups in an organisation so as to enable production and service at the most economic levels which allow full customer satisfaction'. Quality therefore becomes a philosophy and affects all parts of the organisation and supply chain. Quality circles across the organisation are one method employed to deliver the benefits of improvement.

These are not only cost based, but enable employees to deal with issues which have frustrated them and have made them feel as though they may be doing a poor job.

Quality is driven by those closest to the customers and allows rapid feedback. Standards are simple, visual and clear. Simple approaches are adopted to ensure quality problems cannot occur. These are known as poka yoke or fail safe devices, where it makes it impossible for the mistake to be made or to be passed on. These may be incorporated into the design process, where, for example, the components are made so they can only go together in one configuration. In our day-to-day lives, things like not being able to insert a disc into a computer incorrectly would be an example.

Finally, quality systems are introduced. Quality systems as portrayed by BS5750, ISO 9001 and 2, EFQM to name but a few can be very useful if applied to deal with the root causes of quality. Cynically these systems are often imposed and are introduced to play lip service to customers who may have indicated this is a standard required to qualify to undertake business transactions.

At this stage of quality development there is local knowledge about time for investment in improvement. While at previous levels of quality, time wasted due to quality problems was hidden. Once this time investment becomes explicit a need for management direction is required. This then leads us to the evolution of total quality management.

Total quality management

This level of quality is related to the complete system of supply. Not just the production process of a good or service. It is considered that quality is applicable to the whole business and all that the business undertakes. This level of quality looks outside the boundaries of the organisation to its customers and suppliers. At this level of quality, there is a focus on customers. The needs of customers, the processes of the business and the activities which suppliers need to undertake to ensure a robust supply, all become incorporated into the thrust of quality. Here continuous improvement is the responsibility of all those who can influence improvement.

To do this the policy of the organisation, its hierarchy and employees need to be able to deliver quality and hence specifically time is allocated for training, education, team-working across boundaries and employee participation. Organisations refer to a culture of quality and continuous improvement, not as though it is a criticism of those who are part of the organisation, but as though this is an opportunity for the organisation to be better than its competition. This edict is also the driver for supply chains to adopt the same approach.

Total quality management has been heavily criticised for what has been perceived as ‘failure’ as a mechanism for bringing about continuous improvement. Whilst many of the criticisms are well founded, the real issues, which need to be addressed, involve those of management leadership and the ability of management to implement change. Hence the term total quality management may be accredited with failure but the principle of quality is inherently an ingredient for an organisation to be competitive.

There is a tendency in Western management to tick a box when an initiative has been completed, quality requires consistent and continuous review to enable further improvement and change to occur.

Six Sigma

Currently Six Sigma is heralded as the most advanced level of quality (George, 2002). The approach of Six Sigma requires the use of statistical techniques to solve quality issues. There is a premise that all sources of variation can be measured. These statistical techniques are not new and were employed by the quality gurus of the twentieth century. The advance in the use of statistical tools has been made possible by the computer age, where complex mathematical calculations can now be undertaken by individuals trained in the operation of computerised statistical packages.

Six Sigma, like total quality management, is intended to have a whole systems approach. It is popular because it offers bottom line cost improvement on every project. Its appeal is that it uses the language of the executives, such as ROI and ROCE. Charters for projects are designed to establish how much money can be saved through the commitment of resources for the project. To be successful, though, change at all levels of the organisation will be required.

Motorola are accredited with the foundation of the Six Sigma approach. Motorola, in 2003, re-launched their programme following a period of 'taking their eye off the ball' when the approach was not consistently applied across Motorola sites. GE, who adopted the approach around 1995 is probably the only organisation which claims to have a measurement system sufficiently robust to be able to track the savings from projects. The amounts claimed to have been saved by GE and others are vast, and hence Six Sigma has become a widely applied initiative, mainly within large organisations.

The appeal of Six Sigma is its focused and prescriptive approach to solving quality problems. The step-by-step standard approach is based on the plan, do, check, act (PDCA) cycle popularised by Deming in the 1950s and adopted as a founding approach for lean thinking (Womack and Jones, 1996).

This PDCA cycle has been updated from measure, analyse, improve and control and further improved to define, measure, analyse, innovate and improve, control and transfer (DMAI^{2}CT). This has links with approaches used in work study. Additionally, whilst reviewing the design process, a further refinement to Six Sigma has emerged that of define, measure, analyse and verify (DMAV).

Six Sigma is the statistical level of accuracy to denote quality. The measurement of 3.4 parts per million opportunities for defects is another way in which quality is described by the philosophy of Six Sigma. Six Sigma is designed to obtain process knowledge

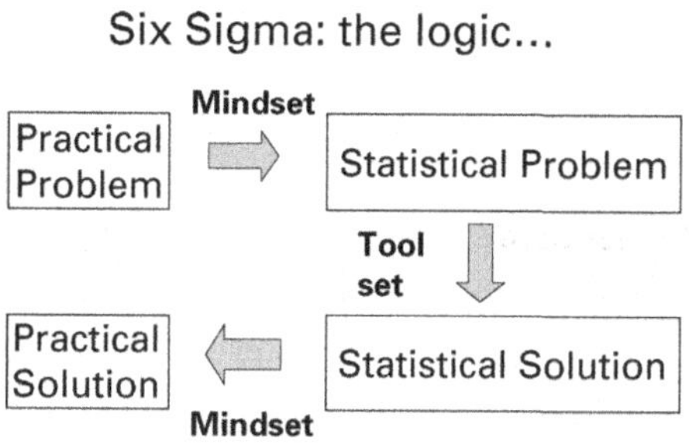

Six Sigma: the logic...

Practical Problem	Mindset ⇨	Statistical Problem
		Tool set ⇩
Practical Solution	⇦ Mindset	Statistical Solution

Figure 6.8 Converting a problem into a statistic to provide a practical solution

which is premised by the view that the inputs to a process need to be controlled to effect the outputs of a process. There is also an assumption that some inputs have a greater effect on the outcome, that is they are more sensitive to change. Finally, in the most complex of situations, the interactions between input variables will make it impossible for simple analysis, like the seven tools of quality, to identify the combining inputs which are responsible for the root causes of problems. In other words the setup of a machine may be affected by a number of control elements. Altering one or a number of these may have a much larger impact than would have been expected. Understanding this relationship is where tools like design of experiment are so powerful for employees, who would have only been able to change one variable at a time without the power of Taguchi's work.

Interestingly, the firm Air Academy Associates claim 80% of problems can be solved by engaging process flowcharts, cause and effect analysis, partition of input variables (constant, noise and experimental) and by using standard operating procedures. This is in contradiction to Shainin's assertions regarding the seven basic tools of quality. We would reiterate all tools have value, it is just deploying them in the right way to the right situation to get the most appropriate benefit.

What organisations need to know is: Can the measurements set for products be met? Is the process capable? Can I process products to meet a turnaround time? How many products produced on this machine will be within the upper and lower specification limits? Can I predict when the process I am using is about to go out of control? All the answers to these questions can be sought by using the tools and techniques of Six Sigma. A word of caution though, there is no point in making a process capable if it adds no value to the end customer!

Variation

A criticism levelled at the lean thinking approach to improvement is that it seemingly ignores the existence of variation. Certainly tools can be seen to take snap shots of a process at any one time and hence cannot be deemed representative of the populations of all data points. There is a philosophical difference between lean thinking and most other

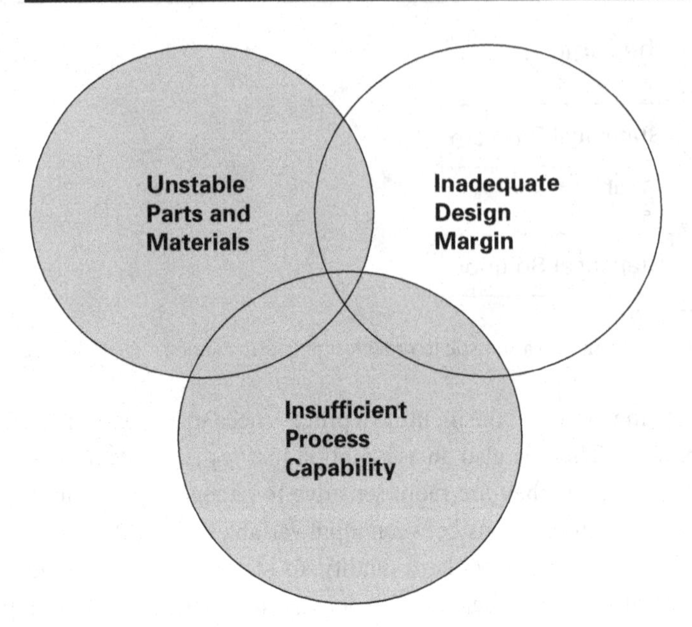

Figure 6.9 Primary source of variation

improvement initiatives. That is common cause variation can and must be eliminated to achieve perfection, whilst for TQM, Six Sigma and theory of constraints it is the control and reduction of variability which is the main focus of attention.

Sources of variation within manufacturing arise from input variables (see figure 6.9). If the materials being used have variability in tolerances or molecular structure, then processing these materials will not eradicate the variability. It could be said that variation is like a virus, it gets passed on, and affects those who come into contact with it.

Insufficient process capability refers to the equipment or line which is being used to manufacture a part. It is not unusual for a manufacturing department to be asked to make components which are at the upper limits of the equipment specification. In this instance, with completely stable parts, the probability of failure can be predicted. If the parts are also unstable, the combination may have catastrophic effects on quality performance. Both parts and capability can be measured and understood. There are tools to help understand and control and manage variability. The final source of variability shown in figure 6.9 is inadequate design margins. For most manufactured products, there are two types of specification limits. The first is the process limits, i.e. what the process is capable of producing, the second is the limits dictated by the design and/or the customer. In the past the latter type of limit has acted as a pass or fail for production. The rich data regarding the range of product output were often not collected (unless SPC – statistical process control – was being used) this resulted in a lack of understanding of the ability to be able to make the product successfully every time. This will be explained further in the next section regarding tolerances.

Quality losses used to be based on deviation from target

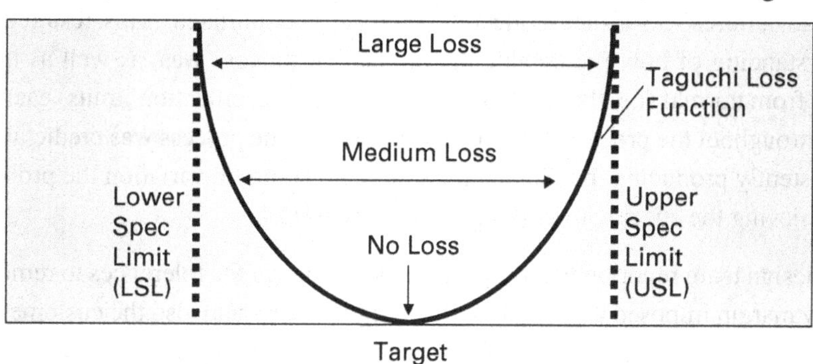

• New philosophy based on conformance to specification

Figure 6.10 New philosophy of quality

Upper and lower specification limits and tolerances

Six Sigma challenges the old view of tolerances and cost. The view was that to make tolerances tight and repeatable meant high investment for a lesser return. This spread of tolerances meant that the root causes for variability in producing a good or service was not understood. The consequence, at its extreme, was that it was luck if you made the product inside the tolerance range or outside. The reason was common cause variation. Predicting the level of rejects (loss) across the range enables control over producing an identical item and this can be achieved through the use of control charts.

Figure 6.10 illustrates the fundamental mind set needed to move from the traditional view of quality, e.g. conformance to specification, to the Six Sigma view of quality, which identifies losses based on deviation from the target. It is counter intuitive because tighter tolerances have traditionally meant higher costs. Once the quality losses from the process are understood, the cost of rejects and rework actually mean that tighter tolerances can be less costly to the firm.

Box 6.3 Steel corporation case

A particular type of steel was required for a downstream production process. The design tolerances for this steel were very tight and the consequence was that a significant proportion of the steel was rejected through the process. On initial review the perception was that the design tolerances were too great and this was not a cost effective product. Two teams were set up to review the design parameters, one with the customer, the second to review the process statistically. The second team started their review by using data which was already available and found that these related only to failures. So a data collection exercise was initiated. This resulted in an understanding of how predictable the production process was, as well as how far away from normal (i.e. the mid point of the process specification limits) each batch was throughout the process. It became apparent that the process was predictable, i.e. consistently producing, but not centred, i.e. with common variation the probability of achieving the specification limits repeatably were low.

The design team reported that they were able to change the tolerances to remove the safety margin imposed as a result of the design process and also the customer could slightly alter their requirements (for the same reason) if tighter tolerances could be achieved in a repeatable manner.

This was proved through a design experiment and the outcome was product rejects reduced by 80% and delivery to schedule improved by 98%. Whilst reject products will be a feature of this process, it is possible to predict this and live with costs and delivery implications.

Measurement systems analysis

If quality depends on statistics, then measurement is an important element of this approach. In a manufacturing environment we see all types of measurement. Things like go/no go gauges, machine operating panels with numbers and gauges and coordinated measurement machines. These measurements are the basis of our problem solving and enable good and bad quality to be detected. However, rarely is it understood that measurement, like everything else, has variability and this is caused by both the product and the measurement itself. Figure 6.11 illustrates the degree to which the component parts of measurement would co-exist, i.e. product variability such as tolerance and measurement variability such as the guage. The statistical approach of Guage R&R analysis enables the proportion of each component to be assessed.

Many measurement systems throughout the production process are not granular enough to detect variability and therefore help identify the root causes of variation and potentially inadequate quality.

As a guide, the following five stages are recommended when starting to understand production measurement systems.

- Stage 1 – Identify the key measures – what is the right information
- Stage 2 – Create standards/definitions for the measures so everyone is looking at the same thing
- Stage 3 – Determine if your measurement system can detect variation

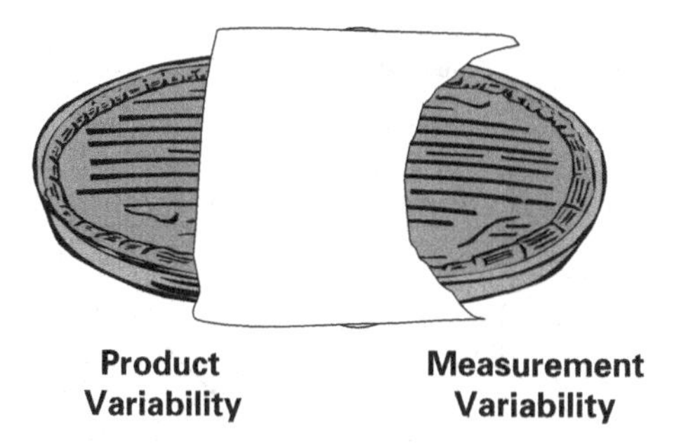

Figure 6.11 Measure systems analysis

- Stage 4 – Decide on method(s) of measurement – 'torturing the data until it confesses' and ensuring that the data process is error-proof
- Stage 5 – Continue improving measurement consistency

The sources of measurement variation are rooted in the actual process and the measurement itself. The process can experience long- and short-term shifts in reliability. In the measurement itself, potential sources can be:

- the variation within the process,
- between the operators or
- between the parts.

These are often distilled further to linearity, stability, repeatability and calibration. What is worth highlighting is that 'skilled' practices such as altering the configuration of the equipment because the operator knows this is 'the right thing to do' could easily be a reaction to common cause variability. This reaction may amplify the variation which is then a special cause and requires a different course of action to remedy. The measurement results being observed may well have sat in the normal range of potential results for this process, but this is lost because of the operator intervention. In this case, altering the equipment settings could end up amplifying the issue by increasing the range the normal distribution of results could take. Additionally the measurement results disguise what is happening.

Similarly, measurement instruments should only be recalibrated when they show special cause evidence of drift. Otherwise, variation could be increased by as much as 40%. This is because adjusting for true common cause variation adds more variation (Deming's rule two of the funnel). To understand about the impact of measurement of performance, the use of the technique gauge R&R is recommended. This is described in more detail in the next section.

How a gauge R&R study works

Statistical tools (such as Minitab) estimate the proportion of the total variation that is due to:

1 **Part-to-part variation:** physical or actual differences in the units being measured

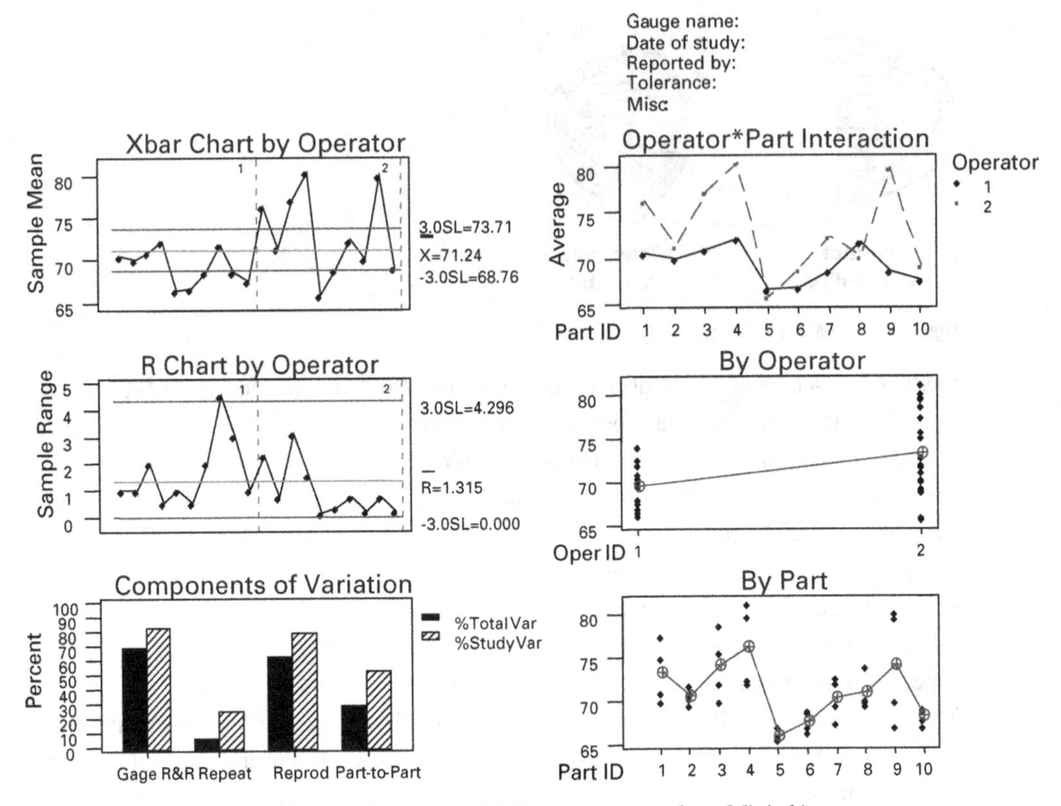

Figure 6.12 ANOVA results for a gauge R&R review (output from Minitab)

2 **Repeatability:** Inconsistency in how a given person takes the measurement (lots of inconsistency = high variation = low repeatability)
3 **Reproducibility:** Inconsistency in how different people take the measurement (lots of inconsistency = high variation = low reproducibility)
4 **Operator–part interaction:** An interaction that causes people to measure different items in different ways (e.g., people of a particular height may have trouble measuring certain parts because of lighting, perspective, etc.)

To understand the capability of the measurement systems which your organisation uses, an exercise to record data from the measurements taken can be analysed using a statistical package and the results would look like figure 6.12.

If there is excessive variation in repeatability or reproducibility (relative to part-to-part variation – see bottom right-hand graph in figure 6.12), you must take action to fix or improve the measurement process. This illustrates the same findings as were being discussed in figure 6.10. The goal is to develop a measurement system that is adequate for your processes needs.

Process:		MGCC Ref stds/controls						
			Importance					
		Importance to MGCC failure	9	1	5	9		
		Importance to rework	5	9	9	9		
Process step name	Process Step No:	Variable	Bias	Outliers	variability	curve shape	Other	**Total Score failure**
get from storage	1	Time	1	1	1	1	1	24
	1	temp	1	1	1	1	1	24
	1	vial to vial difference	3	1	1	1	1	42
Reconsitute f/d	4	volume precision	3	1	3	3	1	**70**
	4	volume bias	3	1	1	1	1	42
	4	water quality	1	1	1	1	1	24
	4	operator	3	1	3	3	1	**70**
	4	time vial out	3	1	3	3	1	**70**
	4	time vial open	3	1	3	3	1	**70**
	4	environment	1	1	1	1	1	24
	4	**cross contamination between vials**	3	3	3	3	1	***72***
	4	**cross contamination from water**	3	3	3	3	1	***72***
	4	Loss of f/d material on opening	3	1	3	3	1	**70**
Thaw liquid	5	time	3	1	3	3	1	**70**
	5	temp	1	1	1	1	1	24
roller mix	6	time	3	1	3	3	1	**70**
	6	speed	3	1	3	3	1	**70**
	6	temperature on mixer	3	1	3	3	1	**70**
	6	loading of mixer	3	1	3	3	1	**70**
pool	9	vial x-contam	3	1	1	3	1	60
	9	tip x-contam	3	1	1	3	1	60
Store overnight	10	temp	1	1	1	1	1	24

Figure 6.13 Prioritisation matrix for final testing

Box 6.4 Medical Devices case

The final testing for a product at Medical Devices was producing significantly variable results. The control charts for the test showed the peaks and troughs very clearly. Also following analysis there seemed to be results which did not conform to a normal distribution (known as outliers). The measurement system was mapped using a simple process flow diagram. This was followed by a cause and effect matrix. This ranked the variables for further investigation and enabled the narrowing of focus to those that were considered most important. A multi vari chart (see figure 6.14), which is one of the seven basic tools of quality, was used to analyse 8–9 days worth of data and three different instruments. The conclusion drawn was that the process was capable within the working day of production, but the capability was not repeated between days and this was not easy to detect without the use of control charts.

Evaluating your measurement system is one thing, actually putting in corrective action is another. Hence using the information gained in this project, a design of experiment was used to understand the impact of multi variables on a process. The design of experiment takes the identified key variables and through use of an experiment identifies

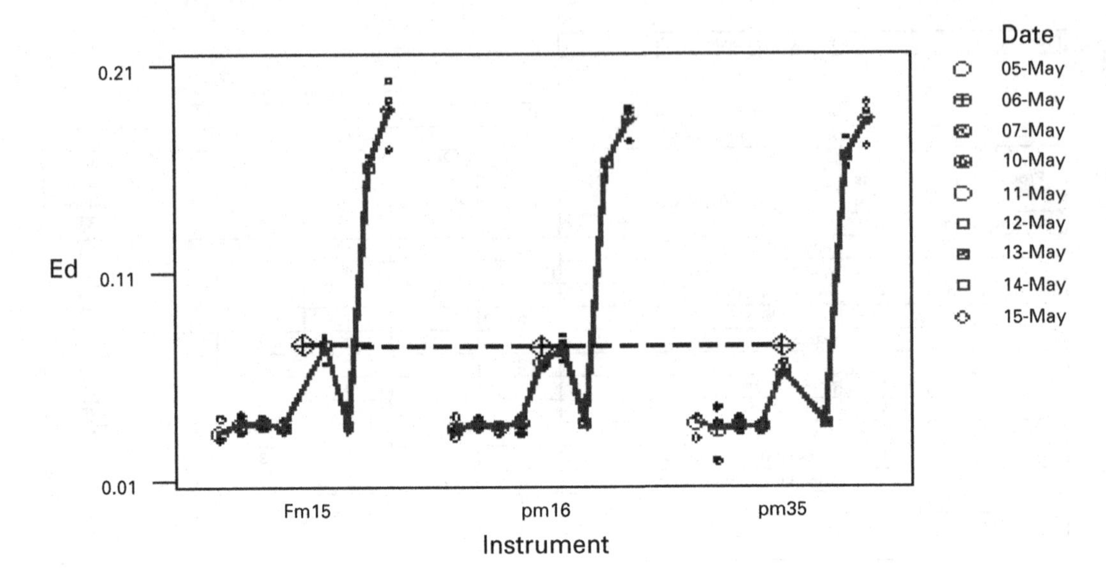

Figure 6.14 A multi-vari chart is a tool identified for use in the seven basic tools of quality (output from Minitab)

the most effective means of combining these variables. This experiment identified that the machine settings used were robust but one machine was behaving in an atypical manner.

This single machine was identified as being the root cause of the outliers and this has meant that good production previously considered poor quality is actually conforming to standards. The consequence is reduced cost and improved adherence to delivery schedules.

There are many other tools which could be illustrated in this chapter, and how they naturally combine with the tools used in a lean transition. So, if what you have read in this chapter has stimulated your interest, then we have added some signposts for good reading and some web sites to further your knowledge. To conclude the final illustration is a whole organisation change combining quality and lean thinking to achieve a high performance organisation.

Box 6.5 Air Repair journey of improvement

Air Repair started on their journey for improvement before 2001. They focused on improvement because of the high level of non-conformance. This in turn was causing a real problem to their customers through the lack of adherence to scheduled delivery dates. Like many organisations which have participated in LERC's learning networks there was a belief that pouring more production orders into the system would eventually produce more output. Instead it caused greater chaos on the shop floor, which in turn pulled more staff, at increasingly higher levels of management, into the task of resolving this issue. As in all organisations, there were also a number of other strategic market and organisational issues bubbling away. Joining the Learn 2 network was one of a number of decisions designed to create access to skill, capabilities and ideas external to the organisation.

As mentioned earlier in this chapter, the issue of quality was highlighted, but was resisted at first because focus on quality had become tarnished with an idea that no change was happening. Kaizen events focused on five day intense projects to establish quick change was preferred along side CANDO (5S or 5C). Figure 6.15 illustrates the process of improvement through a tenacious approach to addressing quality problems. Starting with CANDO enabled action to be felt and seen on the shop floor, but this was not always straightforward, as discussed in chapter 4. CANDO removed some of the opportunities for failure. Overall equipment effectiveness (OEE) was used as an initial measurement of success. After a year it became apparent that to stabilise the production process the root causes of failure had to be understood and eradicated or managed. Measuring variation in the process and the measurement systems became the focus. Statistical analysis enabled the organisation to detect special causes of variation and shifts in performance.

Six Sigma was enabling a greater focus on current improvement because making substantial gains requires, not only control of the production process, but the control of the design process. It is said that performance levels will only reach 4.5 sigma if the design process is not part of the overall improvement strategy. The approach followed by Air Repair delivered dramatic cost savings made as a result of tenacious improvement.

Air Repair was not the only organisation to see significant improvements, Medical Devices and Health Products won national and international awards for their robust systems and rapid improvement in the face of a number of obstacles.

The next stage? Improvement is not considered to be a tick in the box exercise and for Air Repair, the next stage has been to support the strategic supply base to maintain and improve quality and delivery, which in turn will promote an increase in orders. Figure 6.16 shows the proposed cycle for improvement, providing support to enable change where organisations need assistance.

Conclusions

In the beginning of this chapter we posed the question, 'Why quality?' The answer simply is without good quality stability cannot be achieved and abnormalities cannot be detected. Quality approaches in tandem with the tools of lean thinking are a very powerful combination for successful improvement. This is demonstrated by the types of projects and cases illustrated throughout this chapter.

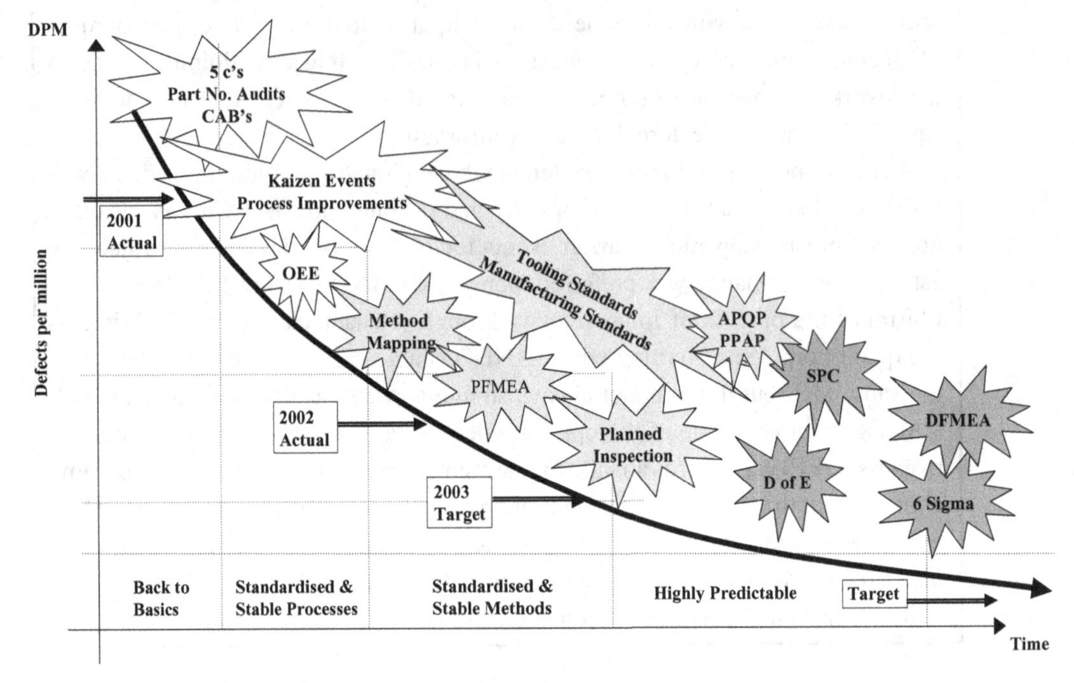

Figure 6.15 Combination of quality and lean activities to improve the cost of quality

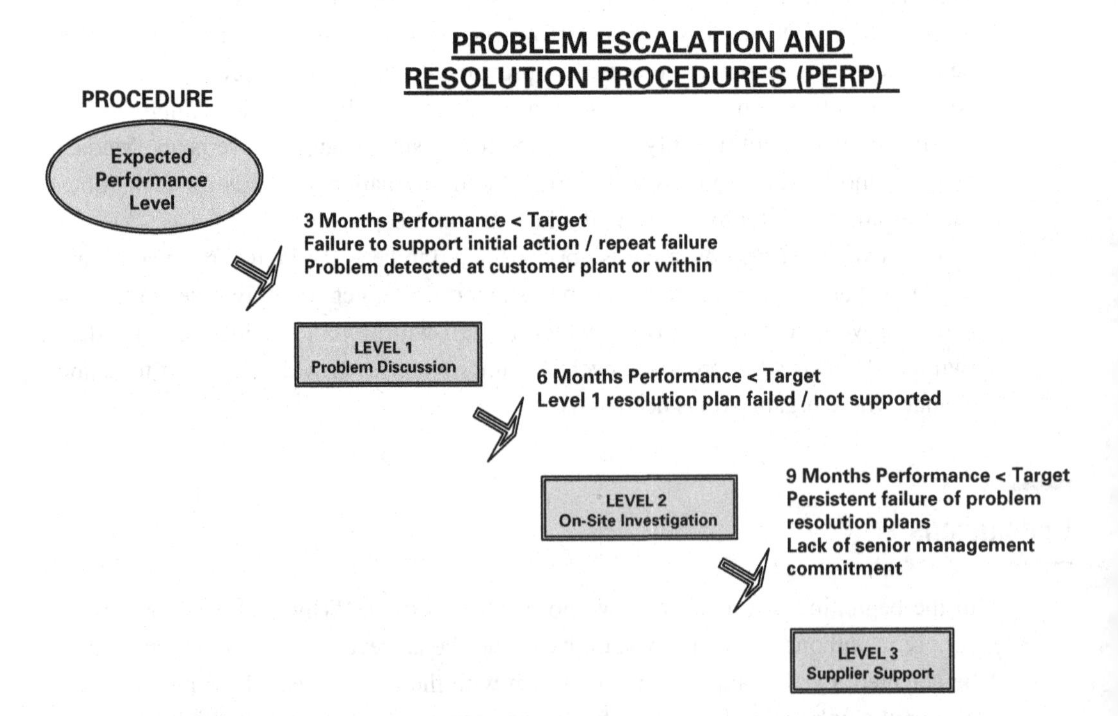

Figure 6.16 Supply chain development and integration

The human capacity to solve problems is an important element of this ability to successfully improve. The combined forces of all the people involved in change enables better decision making. With the statistics and the statistical tools to provide the necessary information on what to improve and why, it is hard to argue why this should not be part of everyday business strategy. Involving all in problem solving not only increases morale but improves everyones understanding of what is really happening to the flow of products being produced for customers.

Organisations can achieve high quality and in particular integrate these levels into an improvement programme. Quality does mean cost reduction and will always need to be linked to the increase of work through new orders or integrating outsourced production. A need to link to key strategic decision making is needed prior to this point. Without this the concerns of redundancy and downsizing could become a reality.

Although various tools of quality were discussed and illustrated, if you want to adopt this approach you will still need to go to a 'how to' book – a number have been recommended below. The sequence and use of some tools which an organisation will need to undertake change have been included. The inference is that statistical capabilities may need to be developed as well as project management skills. Remember quality and improvement are related to the contingent needs of your own organisation. Do what is right for you and the situation you are in. The only rule is learn from what you do and eventually cover all the steps of the journey that have been identified in chapter 2.

Six Sigma is a positive development because it makes the quality tools of the past accessible to us today. Becoming a black belt is not the important step but combining knowledge of this and lean makes your organisation one which will eventually become best in class – a journey which the organisations from the Learn 2 network are comfortable with.

Signposts and recommended reading

A number of books have been mentioned in this text in addition a useful potted history of quality is in the first section of Nick Rich's book *Total Productive Maintenance: The Lean Approach*, which particularly relates to quality and Lean Thinking. For a guide to what the simple quality tools really mean, I often use *The Quality 75* by John Bicheno. I also like the pocket book of *Six Sigma* by Rath and Strong Consulting, again because it gives practical and simple explanations of quality tools and how they can be used.

7 Pull systems

Nick Rich

Introduction

The origin and at the very core of lean enterprise lies the design of the Toyota production system (TPS) and its ability to flow products through a low buffered manufacturing process (Womack *et al.*, 1990). On first encountering the approach to systematically designing a high-performance production system, it appears counter-intuitive to many and a long way away from the Western traditions that have been dominated by mass production thinking (Womack and Jones, 1996). However, the 'system' design combines a number of key features which support customer service and the compression of time between receiving goods and receiving payment for the conversion process (Mather, 1988). This chapter will examine the basic elements of the 'pull' systems of a lean enterprise and explore alternatives for businesses that do not conform to the 'high volume–standard product' environment of automotive component manufacturing and assembly.

Production scheduling

The history of production scheduling in the West began with the use of clerical workers to schedule manually the complex products and the manufacturing/assembly processes so that products were outputted into the finished good stores just as customer due dates for orders were reached. In essence these planners attempted to launch production in time for when the customer expected delivery, but there were many aspects of traditional and mass production approaches which complicated even the most simple of manufacturing processes. These features included a core belief that batch production was the answer to cost efficiency (and therefore profit maximisation). This traditional mantra has been increasingly questioned as the route to achieving high customer service and/or cost efficiency. Nonetheless this approach has prevailed to the modern day.

Upon the advent of the computer, clerical workers were largely 'redundant', due to the new speed at which these reconciliations could be achieved and schedules of

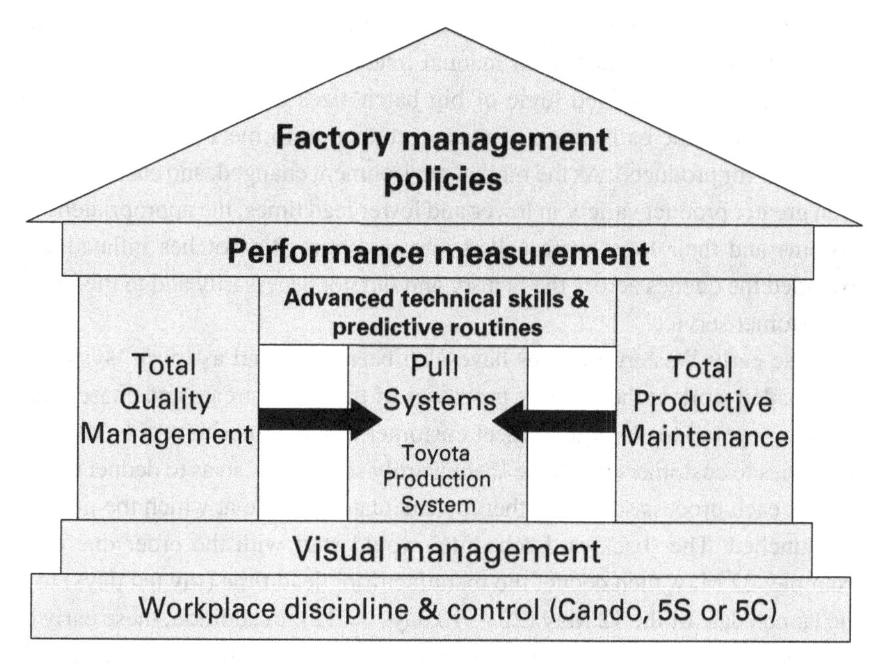

Figure 7.1 The house of lean

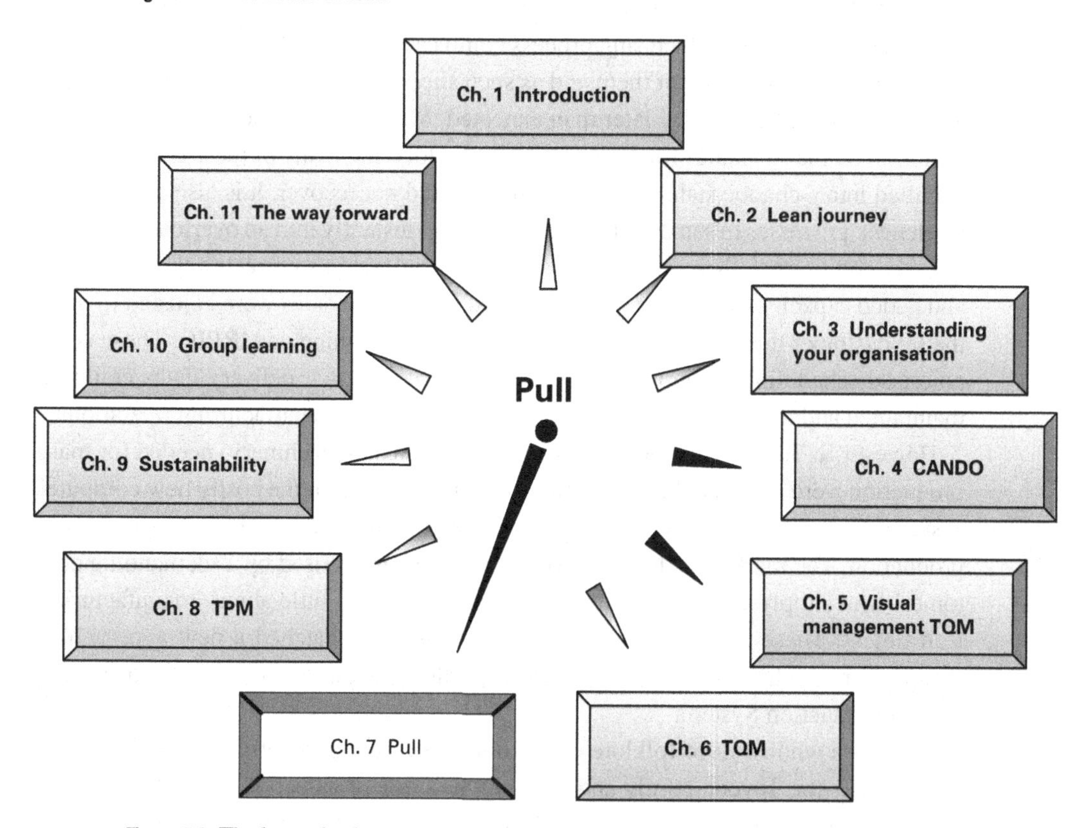

Figure 7.2 The lean wheel

operations printed out. Computerisation brought with it new speed, sophistication and a much lower dependency upon manual intervention and expediting, but it did little to challenge the engrained logic of big batch sizes and high asset utilisation which had dominated the early reign of mass (production to meet an unsophisticated mass consumer for products). As the market environment changed, and customers demanded even greater product variety in lower and lower lead times, the appropriateness of these systems and their logic were called into question. Big batches inflated lead times, extended the queues across the factory and did not necessarily add to the improvement of customer service.

These early Western systems have also been classified as 'push' systems because they loaded work at the primary processes of the value stream and chased them to the finishing operations in time to meet customer orders. To achieve this, the schedule of deliveries to customers would be 'backwardly scheduled' so as to deduct the amount of time at each process stage and thereby calculate the time at which the product should be launched. The 'backward schedule' would start with the order due date on, for example, 22 May, then deduct the manufacturing lead time (say ten days) to establish the launch date of the 12 May (22 – 10 days = 12). In addition, these early computer systems (materials requirements planning or MRPI) were originally 'blind' to and ignorant of the actual capacity of each process stage. As such capacity constraints or 'bottlenecks' in the manufacturing process could cause overloads where they could not process everything needed of them and as such slippages to the schedule would appear as products were made a day later than expected. Missing the schedule therefore compromised all the schedules for materials in the factory and many of these early systems spiralled into a chaotic state from which they would not recover. It is also possible for bottleneck processes to run out of work and then go instantly into an overload situation.

Later additions to these computerised scheduling systems corrected this weakness and added capacity modules to ensure that product launch dates were adjusted to meet the set customer due dates (manufacturing resources planning or MRPII). These computerised scheduling systems scheduled work earlier so that delivery dates could be maintained and these fundamental systems underpin most manufacturing operations.

However, at Toyota in Japan, the resources (capital and machinery) needed for mass production were not available and neither was the funding for the costly new computer systems needed to schedule the complex operations involved in vehicle assembly and production. The market for Japanese vehicles was characterised by a 'demanding customer' who simply would not wait for products and cared little about 'manufacturing lead times'. These market and economic circumstances hatched a new approach to manufacturing which we now know as lean production, but was originally termed the Toyota Production System.

There is a whole raft of folklore concerning the actual origination of the system by its designers the Toyoda family and Taiichi Ohno (the architect of the system). Some authors point to Toyota's visits to Ford in the early 1900s and the 'flow production' of moving vehicle assembly practiced in the USA. Others cite Ohno's visits to US supermarkets and his alleged fascination with how cans of product were replenished when customers selected them and left the store. The true origination and influences of the system design are best left for academic debate – the system itself and its final design is of far more importance and intrigue as a means of high performance manufacturing.

Before we explore the system in more detail it is worth staking out some ground rules that underpin the lean production approach. The first is that, wherever a product can be made within the cycle time of the next machine, then these two operations will be physically linked to feed each other (Womack and Jones, 1996). In the West, the typical factory layout was that of grouping common technologies together to form 'techno-departments'. This did little for the flow of materials – the primary passion of Ohno (1988). Secondly, as the third lean principle cites, where it is not possible to flow production, then a 'pull' system of communication must be utilised. It is here that the Toyota and Western systems diverge considerably. Under the conditions of a pull system, the two manufacturing processes must remain disconnected but (rather than schedule work through the process path) a standard amount of inventory is held between the two. This inventory contains the variety of products that could be needed by the internal customer process and is available instantly. In the West, this inventory was timed to arrive when needed but under the pull system it is available 'on demand' (and known as a 'kanban'). The 'pull' logic is simple (Monden, 1983). With a standard amount of inventory at the finished goods storage, customer orders will be serviced from this stock to meet real demand and actual numbers (not batches!). As this inventory reduces, it will reach certain points whereby the containers of the product are empty and can be returned to the last operation. The return of these items or the cards attached to these boxes (to product finishing area) act as a visible signal that tells the final operation to select the exact same product that has sold and to convert it ready to place it in the finished goods area. Thus products are instantly available for the next time the signal triggers the movement of materials and thereby the physical presence of stock protects customer service. So we now have a loop of communication and production that reflects what has been sold.

As the product finishing area consumes materials, the released boxes (or attached cards) are sent backwards to the internal machining processes to call trigger replenishments of these materials. As such, customer orders pull work through the system and that is the logic of the Toyota production system. Obviously, the inventory throughout the process needs to be managed to ensure that there is enough to keep the customer supplied and enough to allow each operation to work – so production controllers do not look backwards they look forwards to see what demand is expected so as to ensure the inventory is there to allow work to be pulled and flowed between internal customers and suppliers. Taking this to the next dimension, these systems can be replicated with customer stocks (return of cards on consumption at their factory) and also with suppliers (as cards are sent back when raw materials containers are emptied). Now you have

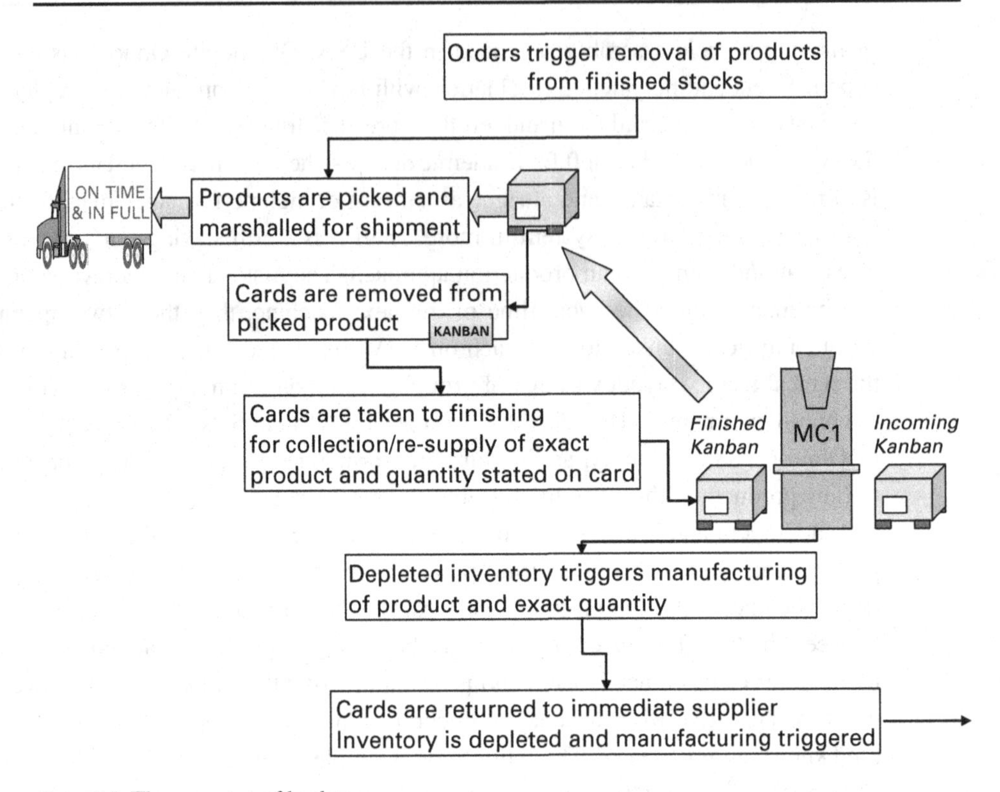

Figure 7.3 The sequence of kanban

the basis for the pull system within the factory and across the entire supply chain. So this is the seemingly counter-intuitive approach developed by Toyota – to displace the vagaries of schedules with stocks which support a very low lead time and a standard of knowing what is selling and what to replenish within and beyond the factory. These standards (container sizes and the way the system is controlled by replenishment of actual sales) is that which creates the disciplined material flow for which Toyota and its suppliers are famous (Lamming, 1993).

The 'pull' approach, described in this manner is not appropriate for all businesses, especially those with high product variety and erratic demand, but it is used here as the stereo-type. The latter circumstances will be reviewed later in this chapter.

The actual words, used by 'lean people' is a complicated vocabulary and this often confuses people when they learn about 'pull' systems, so we will now go through some of the basic terminology used to describe these 'systems'.

The first term used in lean production is that of 'levelling' and this process looks at what orders are being placed for products and attempts to create a pattern for these orders which is 'flat' and allows a small amount of product to enter the stores each working day or even within each hour. The Western approach simply adopted a process of launching batches so the schedules hitting each manufacturing area were 'volatile' and either there was a lot of product to convert or nothing. The pull system uses 'level' and 'flat' scheduling to help replenish work to the stores and to maintain the 'normal' amount of products needed to service customers. The word normal is used intentionally here, as in the batch and push system there is no sense of what is normal – you just process what you get as quickly as you can then pass it on. Under lean production standards are everywhere, including the standard amount of stock in the finished goods storage and the standard quantity in each box.

So the logic of 'pull' is to use these standards to help guide the system and trigger the right replenishments in the right sequence. If orders were not levelled, then everything could be taken at any time and this causes abnormal confusion for the replenishment processes at the different work centres. So with an adequate amount of stock in the finished goods (with safety stocks to cover for customer 'blips' in daily orders), the replenishment rate for all processes can be calculated and also the frequent collection and movement of products can be engaged to 'nibble away' at the kanban stocks throughout the factory and prompt replenishments in a regular and sequenced way.

Obviously, for most manufacturers there will be many different products needed by the finished goods operations to service customers, so a regular route of collections is engaged to ensure that, throughout the working day, small amounts of product in boxes are taken to the warehouse to replenish consumption. Multiple deliveries allow frequent exchange of cards or empty boxes and therefore every process is better aware of what needs to be replenished throughout the day. If there was only one such collection per day, then all product boxes would be removed and none of the operations would know in which sequence to replenish what has been taken (hence chaos). The collection process is known by a number of names but 'the water spider' is probably the best and this refers to the constant movement of collections and returns throughout the factory (Monden, 1983).

The regular collection of standard products in standard containers creates a communication system which is called 'kanban' control. Most typically, each box of standard parts has attached to it a card or ticket (known in Japanese as a 'kanban') and there are different forms of cards. To start the process of pulling work, the final dispatch department take the cards which represent what must be withdrawn from the stores and marshalled for dispatch to the customer. These cards (say coloured white) are then sorted to the load needed for each customer and entered into a levelling box also known as a heijunka board. The levelling box is simply a pigeon hole system which shows when the vehicle will leave and has slots within which kanban cards can be placed. The worker will spread these cards out over the preceding hours to trigger the collection of materials in time for the whole load to be sent out (not issue all the cards at once for collection). So each hour or every 15 minutes, a number of cards are available from the levelling box for collection from the stores or final machining process.

The materials handlers, responsible for collecting products will then visit the box at the relevant time, collect their kanban cards and head off to get the product and the standard quantity written upon the card. These cards are therefore withdrawal kanban

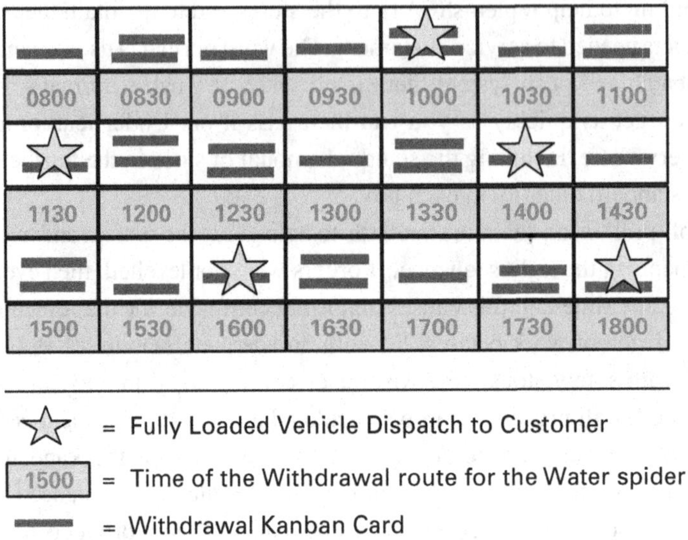

0800 0830 0900 0930 1000 1030 1100

1130 1200 1230 1300 1330 1400 1430

1500 1530 1600 1630 1700 1730 1800

= Fully Loaded Vehicle Dispatch to Customer

1500 = Time of the Withdrawal route for the Water spider

= Withdrawal Kanban Card

Figure 7.4 Kanban levelling box (heijunka)

and give the authorisation to move materials. On reaching the place (written on the card) where they will find the materials needed, they will encounter a standard box containing a standard number of parts. On the box is another kanban card (say coloured blue). They remove this and insert their withdrawal kanban in its place. They then collect the box and return the blue kanban to the team leader in the manufacturing area. This blue kanban is known as a production kanban and is the authorisation to replenish. It never leaves the manufacturing area but cycles within the production area as the withdrawal kanban cycles between the levelling box and the point of collection. The materials handler will then return to the stores and commence the marshalling of the load and later, at the designated time, will go and get the next set of products for the dispatch until the vehicle is ready and leaves on time.

The 'blue' production kanban is kept by the team leader and then sequenced for production to actually happen. To do this, the team leader will take the necessary standardised materials from the kanban buffer between the manufacturing operations and return the card to the preceding team leader in much the same way as before. This instructs the internal supplier that work has left their area and needs to be replenished. The finishing team leader will then produce the required output and locate it in the exact space from which the material handler withdrew the stocks in the beginning. If there is no other card for replenishment, then the work station will remain idle until a collection occurs and so on and so forth throughout the production system, these cards are exchanged and trigger replenishments in the correct sequence and amounts. It is simple yet devastatingly effective as everyone knows what to make and when.

The Japanese exemplar lean companies have become so reliable that these systems work without the need for safety stocks in the kanban (these would traditionally have

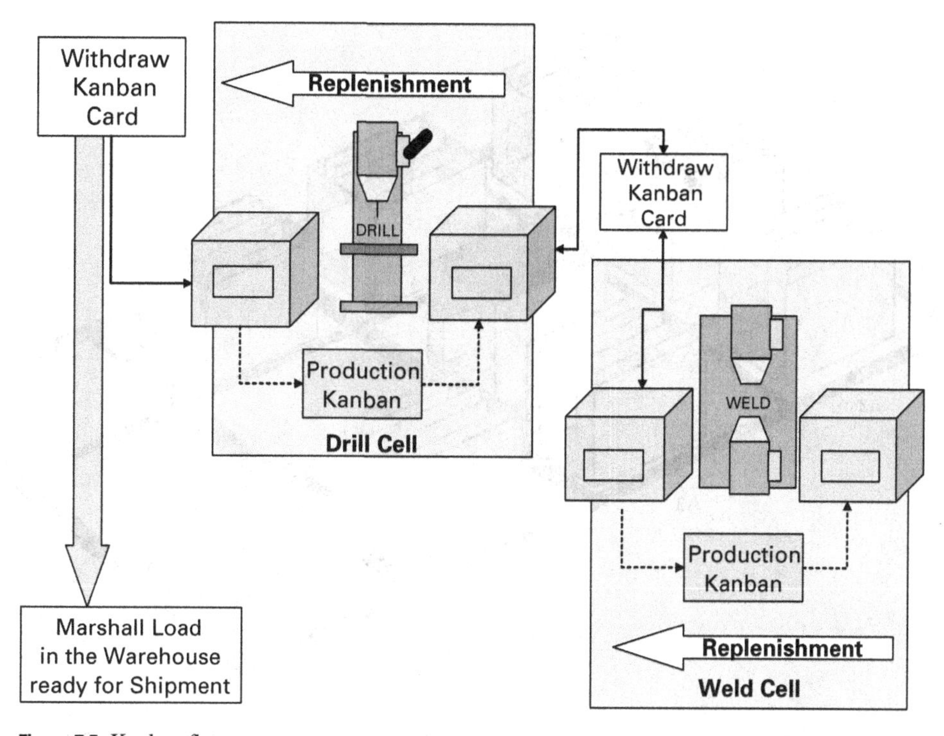

Figure 7.5 Kanban flow

included a number of standard boxes to cover for machine failure etc.). The colour of these cards were 'red' and the difference in colour prompted action because the only reason why a safety kanban would be moved resulted from an unexpected increase in orders or if the internal customer operation was throwing out bad quality products and demanding replenishments for them. Either way, these events could potentially cause chaos, so exposing a red safety kanban meant that the withdrawer would have to refer to the factory management for advice. Therefore abnormal conditions could be easily identified and the system corrected quickly. Many Western companies forget this part of the system and pay heavily for it. Another condition for pull production is that materials held containing safety kanban cards would be rotated, thus avoiding the risk that stagnant safety stocks existed in the system. That is why all kanbans are handled on a first in first out (FIFO) basis and safety kanbans are released by moving the cards around to other boxes and substituting the production card with the safety card. There is a famous case in the UK where an automotive component manufacturing company did not rotate its safety stocks and when they were needed they were found to be obsolete which stopped the Western car assembler for three whole days. The bill, to the supplier, for doing so and demonstrating they had low appreciation of how kanban is meant to work, was enormous!

So this is how the basic pull system works – no real need for interference, just a set of cards and the attention of the production planning department in ensuring the

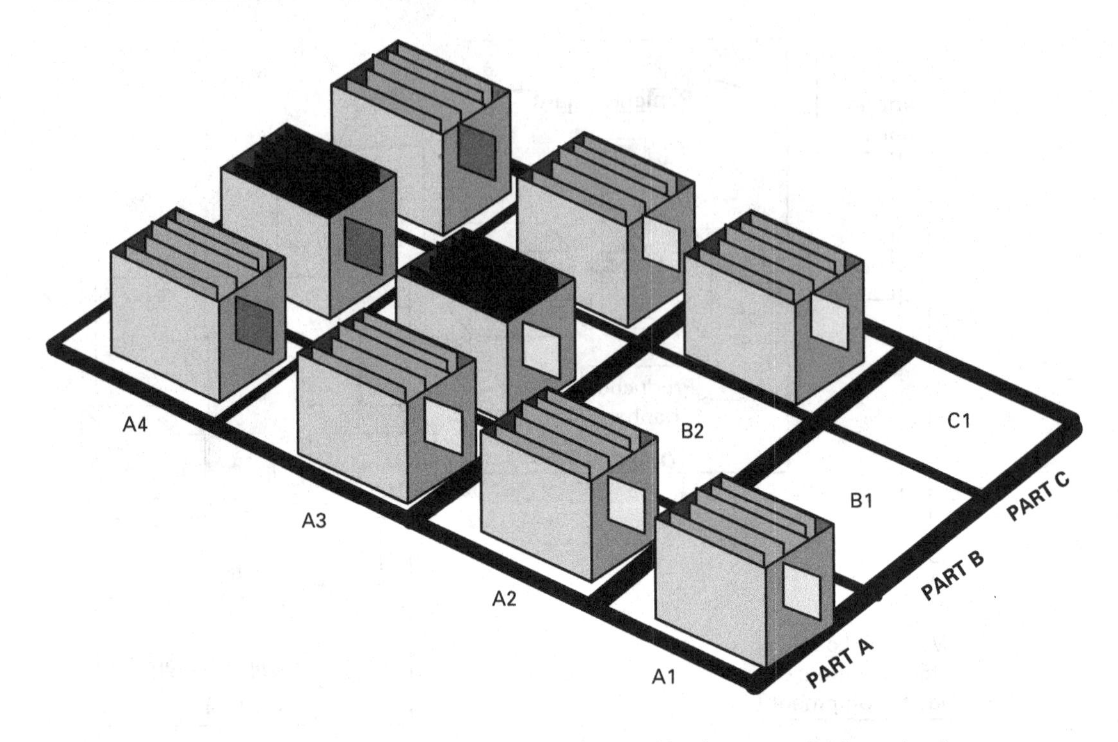

Figure 7.6 Simple kanban floor squares

right coverage of products is in place through monitoring for any uplifts in demand. The other features that support this form of pull production include the establishment of manufacturing cells (Sekine, 1992) so that the core 'running' products can be made and shipped from one location (as opposed to 'special' or low volume orders which would be grouped together in a smaller and more flexible cell or group technology).

In the Toyota system, the overall schedules for the manufacturing and assembly plants is also subject to levelling and this involves adjusting orders to fit the capacity of the factories. The product requirements for each factory are then broadcast each Monday morning to the factories and these schedule requirements show demand in days for the next ten days to form the shipping schedules and then the next few months. Whilst this is called a schedule, it is provided to show what is happening to the order book and whether there is a planned increase in demand which must be accommodated by the assembly and manufacturing operations. This 'directional' information (not a schedule as we know it in the West) is good for the production planning processes at the factory and allows rates to be established for products that are required. In this manner, if a monthly demand has an average of 20,000 units of sales and the factory works 20 days in the month, then the system 'rate' would be set at a replenishment rate of 1,000 units per day. In lean terminology this is the 'takt time' (Monden, 1983) and is the rate at which the factory must re-supply its buffers (kanban) to meet the average rate at which the customer will withdraw products. It is essentially the rhythm of the factory and will be set for a period (usually a month) and then reviewed to ensure that the production system is constantly meeting its customer's demand rate.

In this manner the takt rate can be compared with the cycle time of the production lines, machines and cells to determine where 'pinch points' exist, and, with the benefit of the directional schedule, it is possible to adjust, in plenty of time, the working hours or buffers needed at this point in the system to maintain 'flow'. The concern is always for uninterrupted 'flow' and maintaining the stability of the production system to meet customer demand. There are many other features which could be explored but these are the basics which tie together consumption, replenishment and make processes.

Returning to kanban

As we know, the word 'kanban' means 'ticket' or 'signpost' in Japanese and is used to refer to the 'ticket' that accompanies work in the factory, either during a withdrawal or for manufacturing and on occasion for safety stocks. These are the three generic classifications of kanban and, of all three, management and team attention is spent on projects which remove the need for safety stocks (improved machine reliability or quicker changeovers of machines). To expand upon the rules of the kanban system, there are several points worth noting:

1 One kanban card must be attached to each box.
2 The kanban must state the exact quantity in the container, the location, receiver, part number and any batching requirements beyond the size of the single container and kanban card (i.e. the kanban may be worth only 30 pieces in a box but the batch size is 120 pieces so you must collect four cards (4 times 30 = 120 pieces) to produce the batch of requirements. In this manner you would collect four cards, produce four kanbans and then locate them in the designated kanban area for that product.
3 Consuming department must come and collect the kanban from the production or internal supplier.
4 There is no authorisation to move and withdraw or produce kanban materials without a kanban card.
5 Production must make only the amount stated on the kanban card.
6 Kanban cards must be handled on a 'first in, first out' FIFO basis.
7 Products in kanban must only be placed in the spaces and area written on the card.

The rules of operating a kanban system are very simple, but allow for both standardisation of practice and 'normality' to be observed within the factory. This is however the basic, but most popular, form of kanban. There are many other variants which can be employed to accommodate different types of processes or innovations in communicating replenishment signals. Beyond the simple 'box for box' kanban the next

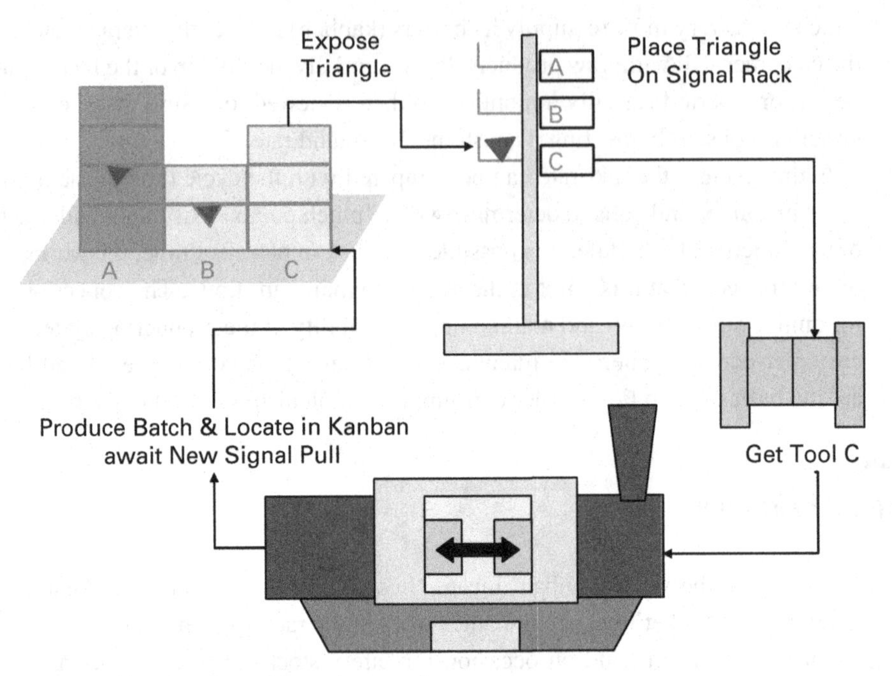

Figure 7.7 Signal kanban (slow cycle control)

major alternative is a system which operates without cards and uses just the physical movement of products from the taped off and colour coded kanban area. This is an appropriate form of kanban for pulling work through a series of desk-based operations rather than exchanging cards (physical movement pulls work through the work stations). A signal kanban is used to control the manufacturing sequence and quantities at machinery with long set ups that necessitate larger batch sizes. Typically these include large die cast machinery producing castings from molten metals. These machines need clear visibility of the stocks being reduced and need a high amount of residual kanban left to maintain material flow whilst tooling and materials are prepared. In this example, the kanban cards are substituted with a metal triangle which hangs from a designated box in a stack of boxes. When this box is removed for consumption, the triangle is handed to the local area team leader and the tooling is prepared ready to produce a large batch to fill up a number of boxes and return the kanban level back to full.

Alternatively a visual board may be used whereby coloured and reversible disks are used to show what materials are in the kanban and what have been taken. Certain disks will have a star on them to show this is the replenishment level and a batch needs to be produced to maintain the kanban levels.

The remaining types of kanban which may be introduced tend to be variants from these basic forms, including the exchange of items that are not tickets but items such as coloured golf balls to designate the product and quantity of replenishments required. It is important not to get too immersed in the means of communication, but to understand

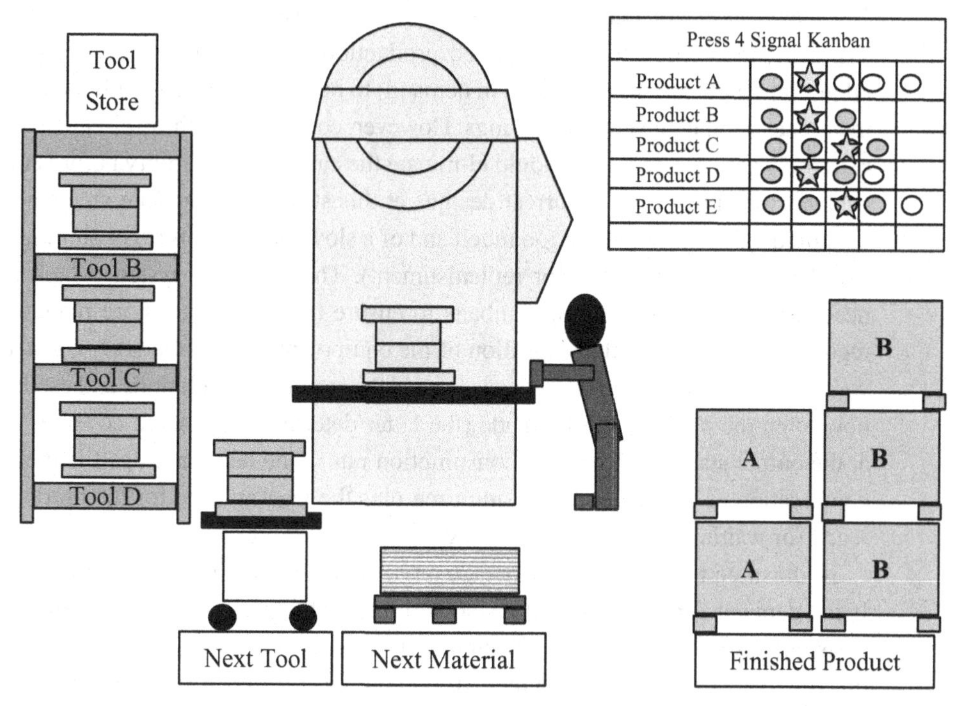

Figure 7.8 Alternative kanban signals

how these systems overcome the basic problem of sequencing and calling for standardised replenishments of items sold to the final customer or to an internal customer.

Calculating kanban requirements

The effectiveness of the pull system is governed by how well the buffer systems are designed to protect against demand, but the size of the kanban is also determined by the frequency with which the product will be made. There is a simple 'rule of thumb' calculation which helps to size kanban quantities.

The number of kanbans required to support a cell or work area is determined by the average demand for the product (items) per day (over a period of time used for planning stability) multiplied by the lead time to prepare and produce the product (in days) plus the safety stock needed (in days). All of this is then divided by the number of items/pieces in a container to yield the number of kanban boxes needed to support the manufacturing process. This final figure is typically rounded up to the nearest box size and then the appropriate coloured kanban can be attached to each box. The next stage in the process is to develop and introduce the necessary monitoring systems for future demand and for the performance of the kanban system to ensure

100% availability of materials for use by internal customers and the final consumer. Typically, the combination of levelled production and the kanban pull mechanisms allows minor deviations (increases in demand) to be accommodated without needing to intervene to inflate inventory holdings. However, companies in these circumstances, of extremely variable demand, should eliminate the causes of variability before engaging kanbans to control flow. Incorrect designs at this stage will therefore create a system of sporadic demand (holding too much and of a slow moving nature) or shortages (and possible chaotic conditions for replenishment). The ideal is a system which has the necessary number of working kanbans to ensure that all products are manufactured regularly and therefore the utilisation of the equipment is kept relatively constant (not waiting or overworked). Again this is largely the result of calculations and establishing how often the product will be made (the latter determines the stock coverage needed at the output stage to satisfy the consumption rate). The lead time itself is therefore a combination of the actual processing time plus the waiting time for the product to be ready (or waiting time between making that particular product).

To illustrate the calculation lets take a typical manufacturer of metal pressed parts. If the demand for parts (over the standard and stable period of one month) is 1,000 parts per day and the lead-time is ten days (the product can be produced and waits for this time between being made) then 10,000 pieces are needed to cover this period. To cover for machine reliability issues, weekend working and other 'noise factors', the company has decided to have a safety stock of three days or 3,000 pieces. The overall amount of stock in the system is therefore 13,000 pieces. To convert these to kanban we need to divide this number by the container size of 750 pieces. That gives us 17.33 boxes which would be rounded to 18. Of these kanban boxes, 14 are needed for regular (non-interrupted production requirements) and the remaining four represent the safety kanban (with the appropriately coloured safety card). This is the basic calculation for the system design and is the foundation upon which more calculations and diagrams of 'how the system operates' can be introduced. Obviously the levels of kanban stocks will fluctuate and then settle down to a repeating pattern of 'normality' from this point on. Many companies struggle during this stage, but it is largely a matter of working through the patterns and cycles of production, whilst ensuring there is enough safety stock to keep the system working during investigations to see why things failed. Also at this point, it is possible to predict the stock turns of the area and the financial cost burden, taken by the company, as the business invests in kanban to protect customer service and flow.

Modifying pull systems

The book *Lean Thinking* (Womack and Jones, 1996) proposes that, where it is not possible to flow materials in a single piece fashion, then a strictly managed 'pull stock' must be introduced to allow the internal lead-time to be broken and offer the ability to 'draw' stock from these points to make up customers' orders. In a relatively stable production system, such as the one for vehicle assembly and component manufacturing, these systems can operate as kanban, but there are many more industry sectors that struggle with the concept of 'pull' and kanban stocks. The first group of businesses involve the 'heavy processing' sector, within which the capital intensity of the production facility is so great and interconnected that it is physically impossible to create kanbans even though they produce only a limited range of stock keeping units. Here it is preferable to look at the finished goods and raw materials stages of production as the pull systems, treating the process itself as merely a form of capital-intensive production line/cell. Basically then the system operates by monitoring what is leaving the business, ensuring there is enough kanban to cover the internal lead-time and a safety requirement, and launching replenishment batches (a bit like a signal kanban).

The other group of companies that struggle include 'jobbing shops' for which there are no such standard products, process routes or dedicated machinery and a second group of companies that have all the semblances of standardised production (machinery) but produce such a range of different products from this standard process that to operate a kanban would be impossible to justify commercially. For example for companies, such as cosmetics manufacturing, it would not be feasible to have a kanban for each colour of lipstick. It is for these companies that another form of pull system needs to be introduced, which combines the scheduling process with a 'fake' kanban. This system is both effective and allows shop floor teams to experience a modified form of pull.

Many of the companies that were studied during the writing of this book fall into these categories, whereby it is not instantly obvious or commercially sensible to set up large numbers of kanban for a wide variety of products/process routes. The modified pull system involves an approach, first developed and then extended by Goldratt (1993), called CONstant Work In Process (CONWIP). Goldratt initially envisioned a manufacturing system with a main buffer located at the bottleneck of the factory or process routing. He logically argued that the bottleneck must be protected as it is the only machine in the entire process which has no ability to 'catch back' production and it is the machine which determines (through its performance and schedule adherence) customer service. As such, no time could be lost at this 'system' constraint. This logic is not the same as the kanban system and instead is termed 'drum-buffer-rope' (DBR).

The DBR system works by ensuring the full availability of materials at the bottleneck so as to lose no productive time through shortages. The control mechanism for this is to establish a buffer at the input stage of the bottleneck. The pace of the system and the need for materials is therefore determined by the 'drum' or bottleneck of the system (basically the rate at which products are converted) and finally a 'pull' mechanism is used to 'call in' more work or to stop the internal suppliers when the buffer has reached its maximum.

Goldratt (1993) argued that his form of production system, based upon constraint management and deliberately limited queues (buffers), would need a schedule to be produced for the 'bottleneck'. These schedules are key and instructed the bottleneck to select the right sequence of jobs to meet customer orders. His advice was not to follow the traditional scheduling process (backward scheduling), but instead to load the bottleneck using forward order information. Some of this thinking has been used by our research in designing an alternative approach to kanban. The major problem with forward scheduling work is that it can extend lead-times as products enter queues at different times and wait for differing periods. So our approach to the high variety approach is to determine the lead time for the finishing of the products, post the bottleneck, and to standardise this time.

Working with the traditional backwards schedulers, it is therefore possible to develop a schedule to replenish finished goods stocks or make products to order. The buffer is still used to protect the bottleneck, but all other areas are backwardly scheduled to show the sequence of production needed but the actual movement of materials is determined by making the buffer look like a kanban. So we tape off spaces at the bottleneck to represent kanban squares. As jobs are taken by the bottleneck, squares become vacant and this triggers the pull to the internal supplier. To pull work, the bottleneck process will look at the schedule and see what jobs are already in the buffer, the empty space is therefore the next job in the scheduled sequence and the materials handlers will go and collect it to fill the kanban space. This is similar to maintaining a constant amount of work in process (using scheduled jobs and a virtual kanban). As one leaves another is brought to the buffer.

Box 7.1 High variety: low volume pull systems

Many of the research sponsors to this programme operate with very high levels of product variety to the point that kanban was not a commercial opportunity. For one such company, the primary process of the product route was to check, weigh materials and then to pass these to a blending operation and from here to a pressing process and on to final assembly. The bottleneck in the process was determined to be the pressing activity in the middle of the process route. The primary check weighing process took only an hour to conduct per job; the blending took eight hours and a single machine, whereas the pressing process took 20 hours. Final assembly and packaging lasted for only five hours in the total cycle of the product. So because the press operation was the bottleneck, the blending and weighing processes could be subordinated to it (they could easily do whatever the presses needed) and the buffer was established between blend and press. The weighing and blending operations were, as lean principles promote, combined. The assembly process could also easily cope with whatever was outputted by the presses so there was no need for a buffer to be introduced here.

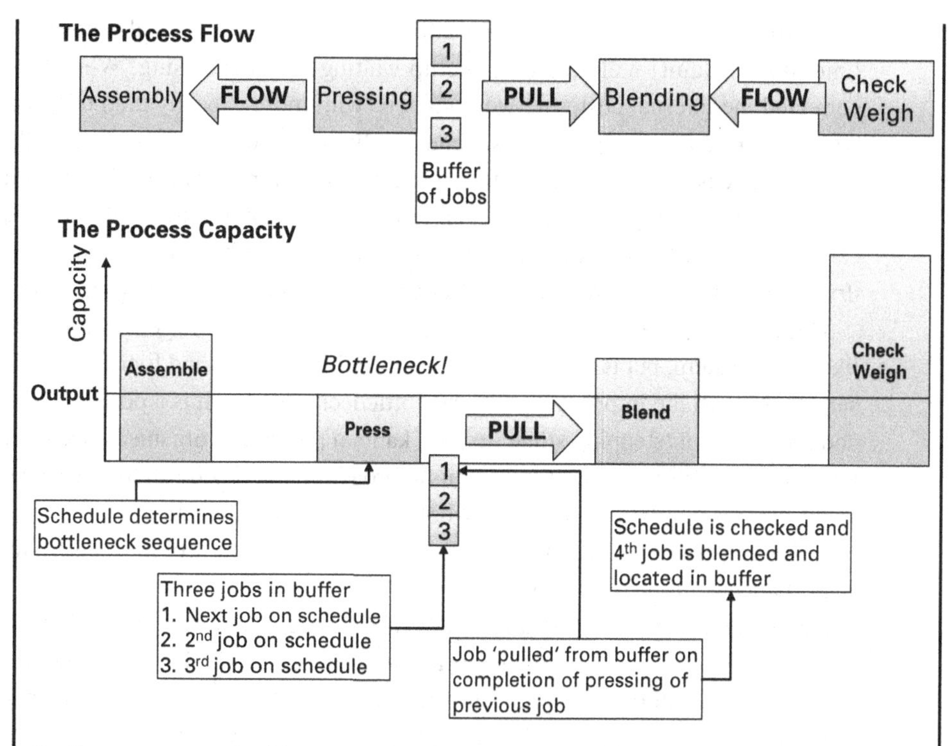

So the system worked by arranging for a number of jobs to be held in the buffer. The number of jobs held in the system was determined to be 3 (a total of 60 hours processing time for the press) and three spaces, representing product pallets were marked out at the output of the blending process so they could see their finished product awaiting pressing. When the pressers took their materials (according to the schedule) they were taking the first in the queue, leaving the second and third to be collected later on finishing the current production. The blenders would then instruct the weighing department to produce the fourth job on the schedule and would place the final blended product in the relevant 'pull' square. Obviously, the system is not pulling work but, for the operations personnel, it felt like the internal customer was triggering the replenishment of work by the blenders (a virtual pull). The three jobs in the 'pull zone' therefore allowed the production planners to change the sequence of jobs by shuffling only the next job and putting it back by one job on the pressing schedule and flow was maintained by not having 'mountains of inventory'.

There are additional benefits of using a 'constant work-in-process' to manage the flow and pull of materials in the factory and this involves the advantage of keeping the bottleneck operation running when there has been a breakdown or failure at an internal customer (a non-bottleneck in the final parts of the process). Under pure kanban rules, the failure at an internal customer work station would halt the pull of products through the system and force the resolution of the problem, within the modified conwip system the bottleneck will keep converting and not stop. Output from the bottleneck (the limiting constraint) will therefore pile up waiting final processing. When the fault is corrected the internal customer work station (with more capacity than the bottleneck) will therefore erode the 'pile' quickly and usually with enough catch-up capability as not to risk final customer service. Furthermore, the most important asset in the factory has not stopped working. Also at the input to the bottleneck, failures (other than quality) can happen and the equipment will still be able to catch up before the buffer runs dry – which in the illustration was 60 hours to correct and catch up production.

This is the logic of 'fake pull' in the high variety manufacturing environment – it feels like a pull system, but is in effect a means of limiting queues and focusing on customer service through the improvement of the bottleneck process. It is a powerful alternative and potentially a 'stepping stone' to pure kanban pull as the organisation sorts out and allocates products to value streams. This approach is also a good alternative when many value streams combine at the factory bottleneck (say an annealing oven or other large capacity-constrained asset). It provides a basic logic and control mechanism to ensure 'pull' happens in the most suitable manner.

Ideally, under the lean and alternative 'pull' system, the ideal batch size would be a single piece, but then we return to the lean principle of 'flow' and not 'pull'. So, even where the alternative approach is taken, it is still preferable to get the batch size as small as possible to avoid clogging up the bottleneck with products which will just 'hang around' in finished goods. The ideal of single piece should never be forgotten though, because it is always preferable to flow rather than 'pull' production. This is not the typical 'law of logic' used by Western managers who have become used to seeing the terms, Just-In-Time, lean thinking and kanban all in the same sentence.

Issues and hints for redesigning your system

It would be wrong to finish this chapter without returning to the process of converting from a push system to a pull system, including what generic sequence of activities are advised and some hints in terms of managing effectively the transition process. The first item of advice to offer is to integrate the personnel in the production planning and control function with the ideals of the new lean system. Under lean production processes, the value adding role of planners is released by stopping them from expediting, and converting them to focus on the medium term (6 months+) capacity management of the firm and to review the levels of kanban needed in the system. Far from removing planners – the lean system reinforces the importance and contribution of such people. It is important to emphasise this at the start and to win the commitment of these valuable employees.

The next set of employees that must be convinced change brings with it a new way of working which reinforces good discipline and compliance are the financial controllers and the quality assurance/compliance personnel. For financial controllers it is important that correct and timely information is captured about the costs of production etc., as well as what products have been received from suppliers who need to be paid. Most information technology systems can handle such transactions and indeed kanban cards can have barcodes added to them to enable scanning of cards and therefore immediate information about material movements and control. Employees with responsibility for compliance with standards set by agencies, such as the Food and Drug Administration (FDA), will need to be concerned with traceability and the ability to audit and trace the manufacture of any given product sampled from any kanban in the system or with historical production records if queries are received from customers. Again such routines can be accommodated by the lean system by recording production activity, the kanban card identifier data and ensuring the kanban are handled on a first-in, first-out basis. During the implementation process, these professionals will need to be integrated to ensure all compliance issues are met and that any production batches can be traced through the total system of manufacturing.

Conclusions

The logic and fundamental elements of the Toyota production system are, despite first impressions, quite easy to understand, and also to assess, how each element of the design reinforces all others. This is the cleverness of the system and how it maintains its steady state material flow within which even small deviations away from the standard (workplace organisation, kanban quantities etc) can be detected easily. The true art of getting these systems to work effectively and efficiently lies in the hands of the production planners. The role of these individuals is value adding when they are monitoring the system and is not involved with expediting work or constantly interfering with production (cutting batches short, closing orders early or constantly rescheduling). These individuals are also those who tend to fear the implementation of these pull systems the most, due to a fear that the new systems will fail and the planner will be back fire-fighting again. The only true answer to this 'vicious circle' is to work through the mathematics of each work station that needs a pull system and to calculate the safety stocks needed to protect it. The planner can then rest assured that they have invested some of the company's money in 'slower than ideal' moving stocks which can be reduced over time as the system stabilises.

An effective kanban pull system therefore allows the people in the process to control the process, it allows planners to plan, and it creates a steady flow of materials in a controlled and simple manner. It is important to remember the use of the safety stocks and also that you should not pull products when cycle times are close enough together for internal operations to simply feed each other. Too many businesses think they must have a kanban to show customers and visitors as this is a visual sign of lean manufacturing, but this has missed the point if the process is capable of flow!

Here lies the final point to be made about 'pull' systems, value stream maps will identify the points in the production system where a pull buffer is needed, but to introduce such a system requires careful planning. Pull systems are advanced techniques which can easy go terribly wrong and stop rather than enhance material flow. To engage 'pull systems', managers will have to investigate process reliability and quality performance of the internal supplying department/cell. Supplied defects in the kanban system will be 'thrown out' and replenishments ordered (as cards are returned) and this creates a sense of crisis. On the other side of the equation, erratic swings in forecasts and actual demand will create swings in production requirements and will make levelling difficult. Another issue which managers will face is how often they intend the manufacturing stations to repeat the cycle of making products (ideally the shorter the time between repeating production the better). Long durations between 'repeats' will have the consequence of inflating kanban stocks (as a result of increasing product lead time). These issues require 'design' and it is the responsibility of managers to do this. Getting these calculations wrong or not spending enough time preparing or improving the process to the point where kanban can be introduced is the major cause of failure. Getting to grips with these issues will create a standardised, stabilised and effective material flow system and the bedrock upon which customary practice can be continuously improved, changeover time reduced, unnecessary safety stocks removed and workers aligned with the real demand for products. When correctly designed and maintained, the kanban system is highly effective and the management of on-going kanban levels can be deployed to production teams rather than the traditional legions of clerical expeditors that chase products in the factories of the past. Ridding production controllers of these reactive tasks and focusing them on the medium-term capacity and kanban levels marks a fundamental shift to getting this indirect function to add real value to the business!

Signposts and recommended reading

My favourite books on the Toyota Production System were written by Professor Monden (1983, 1993) and are highly recommended as insights into the system design. Additionally, Shingo (1988, 1989), Suzaki (1987), Lu (1986) and Louis (1997) explore the dimensions of the kanban system. For a simple read also try Gross and McInnis (2003). Alternative systems to that of kanban, and their supporting structures, are well described by Goldratt (1984, 1986, 1991, 1986).

8 Total productive manufacturing (TPM)

Nick Rich

Introduction

The maintenance function and its relationship with production operations is often a neglected area of lean production systems design, yet the skills of this department, when harnessed correctly, add a new dimension to the competitive arsenal of the firm (Willmott and McCarthy, 2000; McCarthy and Rich, 2004). In truth, maintenance engineering and its associated skills, has been neglected for many years. However, as early as 1970, the exemplar Just-In-Time (JIT) supplier companies to the 'world class' Toyota Motor Corporation had already spotted and were closing this 'missing link' of sustainable high productivity management and had begun to 'blur the edges' between the operations and maintenance functions so as to integrate both departmental improvement efforts. These businesses included the Aisin corporation and the world renowned Denso (then known as Nippondenso) and the approach was demonstrated to be so powerful that it has been promoted by the Japanese Institute of Plant Maintenance ever since (Nakajima, 1988). These early pioneers, Denso and Aisin, were quick to understand that with a lean production system of no stockpiles of inventory and shop floor problem-solving groups, the performance of the factory maintenance engineering needed to be improved. These businesses therefore set in process a chain of events which would later lead to the development of the total productive maintenance (TPM) approach to high performance manufacturing or total productive manufacturing as it is now more popularly known.

Despite including the term 'maintenance', TPM is far from a functionally driven maintenance improvement initiative (Rich, 1999). It is instead a company-wide approach to improvement that overcomes many of the 'sustainability' problems associated with the typical interpretation of shop floor groups looking for improvements. It is perhaps the term 'Total' that is missed by most Western businesses that are quicker to focus upon the term maintenance and assign it to a departmental initiative where it lacks the transformational power of a strategic and cross-functional improvement initiative.

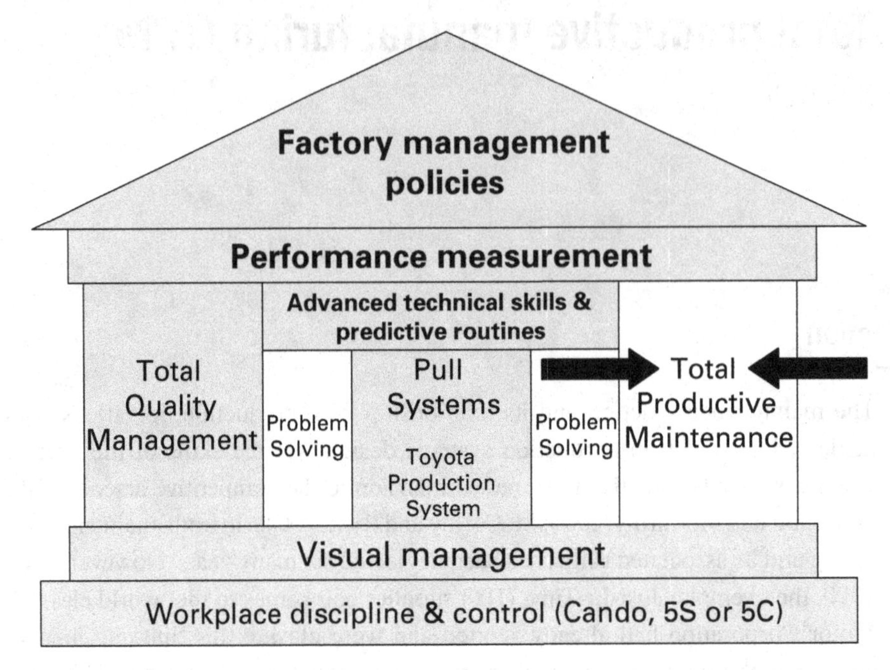

Figure 8.1 House of lean

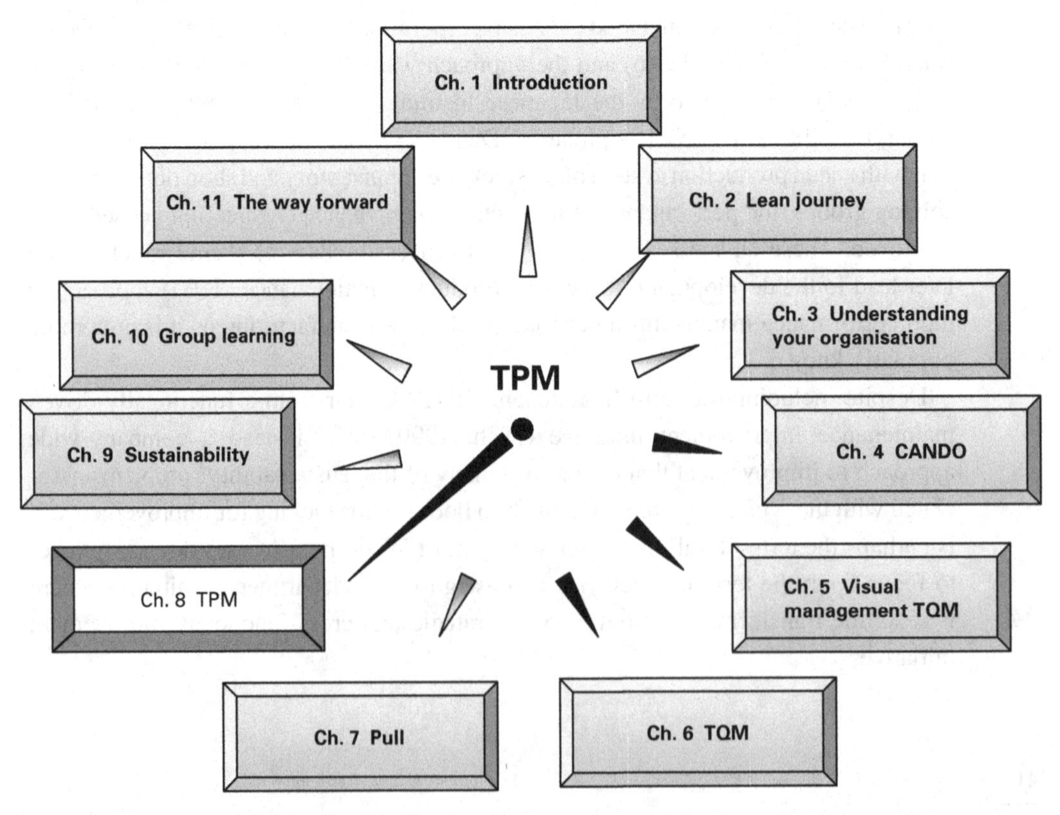

Figure 8.2 The lean wheel

To cut to the point, no journey to 'world class' performance is complete without TPM and no 'world class' business can ever hope to succeed in managing a 'zero loss' and 'zero waste' facility without a functioning TPM programme (McCarthy and Rich, 2004). The requirement for TPM is easy to identify: with flow processes and a low buffered production system the major risk of failure comes from asset breakdown and thus the need to maintain this equipment to the highest standard. Failure to maintain assets at this level therefore exposes the entire business to the disruptions of breakdowns and, with no inventory in place to buffer out these interruptions to product flow, the introduction of the TPM programme has a dramatic and direct impact on customer service. There is another point here too. Most Western businesses work with line operators who know little about the functionality of the machines they control during the normal working shift. Furthermore, without technical training, these operators will inevitably hit the 'glass ceiling' of problem solving, whereby endless problem-solving meetings will highlight issues with the machinery, but, due to a lack of technical knowledge, the team will not be able to solve them (Rich, 2002).

Erratic machine performance and interrupted product flow is not the 'lean way' – the lean approach demands that any abnormality in the workplace attracts instant attention. In this manner, process problems can be identified and corrected quickly and then practices put in place to prevent recurrence. Thus erratic performance of assets – an asset can cost as much as the typical house owned by a line worker – cannot be left to continue in today's stockless lean production systems. So a new approach must be found, which supports the lean objectives of the firm, directly addresses the optimisation of the technology, and 'up-skills' the technical competence and diagnostic skills of line workers (Womack and Jones, 1996). If the new business design does not contain this new blend of skills and processes, then line workers will quickly become demoralised as endless rounds of 'cause and effect' analyses highlights machine-related problems that the teams themselves lack the skills to solve. In this manner, the lean TPM approach is not easy and demands the 'total' integration of all managers that directly interface with product flow – not just the Chief Engineer and the crews of technical staff that interact with teams on a daily basis.

To broaden its appeal, and to lose the traditional emphasis on the word 'maintenance', the TPM approach has recently, in the West, been termed total productive manufacturing (a more reflective name) intended to broaden TPM's appeal to the wider body of factory managers who each must support the programme. The name of the approach is semantic and more an issue of 'internal marketing' than of any real deviation in the methodology and techniques employed. TPM is a means of controlling the efficiency and effectiveness of the production system and it is a feature of all advanced 'lean thinking' organisations witnessed by the on-going dedication of resources allocated to this initiative (over 30 years since beginning TPM, the pioneering companies are still chasing the last vestiges of waste from their systems).

Silver bullets, initiative fatigue and fashionable management

Throughout the 1980s and 1990s many management books and academic publications heralded new business models that would, if applied correctly, radically transform the firm into a 'world beater', capable of meeting the demands of the market and fending off competitors. Also, during the 1980s, Japanese texts explaining certain manufacturing techniques of high performing firms were translated into English, creating an interest in applying these techniques within Western workplaces. Many of these models were however to prove seductively rational but without a methodology for implementation (Rich, 2002). Some of the techniques provided a methodology but offered no real advice concerning how to integrate them into business-wide improvement activities beyond the department that initiated the improvement. Both types of programme however were readily adopted by Western managers. These managers were motivated by a number of reasons: some were keen to implement and be seen to be at the 'cutting edge', whilst others grasped at these new practices as if they were lifelines and implemented change with an air of desperation seeking any form of improvement as a goal.

Whilst improvements were implemented and sustained by some Western firms, many were not and ended in failure – this has traditionally been the outcome of TPM initiatives. It is not an easy process and it calls for the integration and support of many managers. These failures to create a robust production system did little to increase the credibility of managers with employees and even customers. So do not be fooled – implementing TPM will be regarded as 'just another initiative' by many people until these same people are engaged properly, trained and developed to support TPM. The apathy comes from the history of failed initiatives that have haunted their memories to date.

Traditional change programmes failed for a number of reasons, which, for the most part, included a restricted focus of the improvement activities to a series of piecemeal activities, departmental improvements or just 'technical quick fixes' to get over a production problem. Examples of these programmes included attempts to compress the set up time of machines in a belief that the company would be capable of producing high levels of variety as a result. Despite the heralded savings very little tended to be added to the bottom-line profit or customer service offered by the firm. It was therefore a pity that no one included the production control department in the initiative as they have the authority to change production batch sizes and to withdraw inventory from the finished goods held by the company because of the long changeovers. The failure to exploit these 'point activities' is unsurprising, as most Western firms do not operate their production systems as a cross-functional management task, spend too little time involving all the key stakeholders in the change process and pay little attention to standardising activities. The sustainability of improvement is therefore the result of luck rather than management – again a practice that is 'frowned upon' by lean companies.

For those businesses that elect to follow the TPM path, there is no such desire to engage in the practices of old, that have yielded so little, no way to pay only lip service to this important change programme, because each element and pillar of TPM is designed to expose weaknesses in the overall system, and, finally, there is no escaping the overall equipment effectiveness (OEE) measure that governs the progress of the TPM programme (Nakajima, 1988). Furthermore, an effective TPM change initiative will bring benefits to all who engage with it. For managers, it brings a new set of processes and measures, for trade unions it brings a greater involvement in workplace decision making and new roles, and for the customer base it brings 'dependability', reliability of lead times and 'zero loss' products. Most importantly, for technicians it releases time for value-adding project work, which pushes the diagnostic skills of the technician to the limit, and for line operators it enlarges and enriches skills and moves the worker from 'just a pair of hands' to a true business asset upon whom the business is dependent to sustain improvement in the workplace. Again don't be fooled by what TPM has to offer, it is not an easy option and, across Europe, only a few exemplars exist.

TPM definition

TPM has many different titles, for some (the 'purists'), it is total productive maintenance and inextricably linked to the excellent work of the Japanese Institute of Plant Maintenance and its many experts. But even within the purists there are different schools of thought and ways of implementation upon which there is disagreement upon which is best and in which manufacturing environment. Whichever 'school' you care to follow, the type of activities involved with this programme are the same. It is 'total' because it is a company-wide approach (led by senior management and involving every employee) to improving the lifecycle of machine performance (from design to disposal and from operation to diagnostics). It is 'productive' because the objective of the programme is to increase the value added of asset-based improvements for commercial exploitation – yes, turning improvements into something that the business values and can sell to someone else (like freeing capacity and inventing new products). Finally, it is also about the 'maintenance' or 'manufacturing' and this concerns the key skills, rather than our Western stereotypes, and reflects a bias towards engineering knowledge applied to make more products, make them better, faster, safer, increase the variety of what can be produced in a single shift and do all this at a much reduced cost. These may appear high goals, but they have been achieved and improved upon by most TPM exemplar organisations. This is why the TPM approach is so powerful and much more potent than traditional tools and techniques applied in the factory.

These definitions give you a flavour of what TPM is designed to achieve, but in overall terms TPM is quite difficult to fully define in a way that is concise. So before progressing any further, we will look to the cited objectives of TPM for some clues.

Box 8.1 TPM definitions

TPM is an evolutionary approach to excellence in maintenance which aims to eliminate breakdowns by use of the full range of maintenance and housekeeping techniques. TPM builds up the role of the operators and of the maintenance engineers. (Harrison, 1992)

It's an 'OK' definition but does not give full justice to what TPM actually is – His definition is too narrow and too focused on the visual elements of TPM which represent only a small percentage of what TPM businesses get involved with.

A company-wide approach to the management and operation of all the factory assets, both human and equipment, in such a manner as to achieve the optimisation of the conversion process and the generation of customer 'value' over the economic working lifetime of the assets employed (Womack and Jones, 1996).

A better definition which covers the major elements of TPM and identifies the key points of focus that will drive the implementation plan.

The objectives of TPM

There are a multitude of TPM objectives and these do not vary by business (they are just more or less important but still important). The main aims have been summarised in 1988 by Nakajima (the Godfather of TPM) as:

1 To achieve and sustain the most effective use of equipment,
2 To bring together all the people involved with machinery and its planning,
3 To create a robust system of planned maintenance and small group improvement activities working as autonomous groups,
4 That are duly supported and led by senior factory management.

He went further to outline the six objectives of the approach which are summarised as:

1 to reduce downtime from all sources,
2 to reduce variation in performance,
3 to extend the life of equipment,
4 to prevent major equipment repairs,
5 to improve process capability and, finally,
6 to improve the flexibility of machines.

In short, the classic TPM objectives to enhance the quality performance of the firm, to improve the dependability of the delivery process and to reduce costs necessitating a 'total buy in' to the approach if it is to yield sustainable and commercial success. The Nakajima view of TPM does also provide us with some insight into the breadth and depth of implementation needed and takes us well beyond the simpler definitions to a powerful change programme that is factory wide and can then be extended to suppliers.

Lean production, total quality management (recently the Six Sigma approach) and TPM therefore share a common DNA. These approaches all advocate the elimination of waste and variation as a means of improving the speed, quality, dependability, capability, costs and flexibility of processes, and, not to forget, of extending the lives of machines so that they earn money over many more years than the value given in the books. For businesses on a lean journey, TPM is a means of addressing the deadly wastes that plague most factories and the seven wastes that form the Toyota mantra of improvement targets.

The seven wastes represent some of the improvement areas of TPM, but they are not the primary targets. These targets include unplanned downtime to be reduced to zero losses, defects to be eliminated to the zero defect level of production and the performance of the assets to be improved such that the equipment achieves a zero speed loss performance. These three elements, which govern the TPM approach, will be revisited later in the chapter when the overall equipment effectiveness (OEE) measure will be discussed in more detail.

So let us return to the definition of TPM and add a few more. Each of these will provide more insight into what will emerge from the implementation process. Peter Willmott, a friend, defines TPM as 'Maintaining and improving the integrity of our production systems through the machines, equipment, processes, and employees that add value' (Willmott and McCarthy, 2000). Hartmann adds, to 'permanently improve the overall effectiveness of the equipment with the active involvement of operators' (1992) and Maggard (1992) suggests, 'TPM is a new work system that addresses the interface problems between a company's maintenance organisation and its production organisation.' Then there's Masaaki Imai (the guru of continuous improvement [kaizen] techniques), who offers us, 'TPM aims at maximising equipment effectiveness with a total system of preventative maintenance covering the entire life of the equipment. Involving everyone in all departments and at all levels, it motivates people for plant maintenance through small-group and voluntary activities' (1986). Here is the last definition, 'A company-wide approach to the management and operation of all the factory assets, both human and equipment, in such a manner as to achieve the optimisation of the conversion process and the generation of customer value over the economic working lifetime of the assets employed' (Rich, 1999). In many respects these definitions are really defined as goals or outcomes, but overall there is a lot of similarity and agreement of the major facets of the approach.

The totality of TPM

Having looked at these various definitions it is now time to return to the 'totality' of TPM programmes and to explain what this actually means in practice. The term 'total' embraces the following aspects of operational effectiveness of the change programme:

Table 8.1 *Toyota seven wastes and TPM*

Waste	Illustration	TPM Antidote
Over production	Large batches launched against forecasts for stock.	Removal of need to produce in large batches by addressing asset reliability, improving process quality, improving asset speed and reducing set up times.
Unnecessary inventory	High work-in-process and finished goods stocks in excess of demand. Holding materials 'just in case' a problem happens!	Reduced requirement to hold stocks to cover for poor asset availability and quality.
Inappropriate processing	Using very complicated and sophisticated machinery to conduct simple routine tasks which would be better performed by less sophisticated equipment.	The constant analysis of asset performance to discover fundamental weaknesses in asset design is used, through a process called early equipment management (EEM), to design and acquire machinery that is 'fit for purpose' and specified to meet the demands of it.
Unnecessary transportation	Long travel distances between processing stations including potential for product damage.	TPM uses analysis tools to find the causes of downtime (material not available) and it focuses on bringing processes closer together under workplace organisation reviews.
Unnecessary delay	Materials just hanging around the production areas waiting to be processed.	Workplace organisation restricts the amount of materials that can be brought to an area (see Cando chapter).
Unnecessary defects	Lost time never reclaimed and worse still having to pay to clear defects that cannot be sold.	The quality pillar is directed towards improving performance by eliminating all human and equipment related sources of defects/yield loss.
Unnecessary motion	Ergonomics and poorly designed processes resulting in operator stress and fatigue.	The safety pillar is used to assess the ergonomics of the factory and combines with the early equipment management pillar to yield machine environments that are stress-free and comfortable for workers.

1 A total approach to the maintenance of the asset from specification to commissioning and then throughout its life (even when it has been depreciated to zero book value but still process capable and capable of earning money for the business).
2 A total approach to the organisation of the TPM programme involving senior directors (awareness, steer TPM pillars and review performance), middle managers to co-ordinate and drive the process and then execution of the improvements by teams (multi-functional).

3 Total involvement of machine suppliers in designing robust production systems with 'vertical start-up' and peak performance from the first day of operation.
4 Total employee development involving all employees in learning new skills, forming an effective bond with the workers so skills will be passed [with the right level of competency and quality assurance] from maintainer to operator and between the indirect departments and the operations. These skills include interpersonal skills, technical skills and also investments made by the business to blur the formal demarcation between workers to enable an easier flow of cross-training to occur.

So TPM is as broad as TQM activities and lean programmes, it is a major support to the business of high performance manufacturing and, in many ways, the missing link in the puzzle for 'world class' productivity. It is also a project that is not for the faint-hearted and, on many occasions, you will face uncomfortable times of near-rebellion and you will have to win several battles concerning the comfort of employees doing jobs they have known for years.

Productive means eliminating the six big losses/wastes

Now if you take any process in a factory there are three areas of performance which suffer from losses or, to put it in TPM terms, three areas of the factory performance which must be improved to the point of 'zero loss'. Every improvement towards a zero loss scenario is therefore a release of productive power that is capable of commercialisation either by lowering costs or by creating additional capacity for sale. The 'zero losses' are quite logical and the first two include a zero loss of availability of the production equipment and the second is a zero loss of quality performance. Finally there is a 'zero loss' of speed of the equipment relative to how long it should ideally take to manufacture a product. Whilst most TPM books do not split these into two sections, I have done so because the first two activities are of prime importance and the achievement of 'zero speed loss' is a more difficult challenge requiring a much greater attention to detail and sophistication of data collection. The purpose of eliminating the losses is therefore to create a perfect production performance every time the line is scheduled to work and also to withdraw unnecessary safety stock from the factory as it is no longer needed to cover for erratic production performance.

These measurement areas (availability, quality and performance) form the basis of the infamous overall equipment effectiveness (OEE) measure and it is difficult to find a measure that is more exacting or interesting as a means of focusing improvement activities. Common mythology has also deemed that an 85% OEE rating is the threshold of 'world class'. However this figure was bandied around in the literature during the 1980s and things have moved on since then. Our own research would suggest the 'world class' threshold is now much closer to 94% OEE rating.

Table 8.2 *Involvement and roles*

Grade of employee	Inputs to TPM process	The purpose
Senior managers	Chair key TPM pillars and set out the business case. Review performance and herald success.	Guide the pilot and roll out of the TPM system with clear goals in terms of exploiting any gains and growing the business.
Operations managers	Personnel, time and endorsing the programme publicly.	The greatest number of organisational employees work for operations and must be involved. TPM must be seen as both a strategic initiative and one led by operations.
Production control	Allocating time in schedules for planned maintenance and improvement work.	To engage production controllers in such a way that they get more productivity as a result of committing to schedules that are not maximised in the short term. The big gain is in the accuracy and reliability of the resultant production system to be scheduled accurately in the future.
Quality assurance	Standards and calibration of the production line tasks.	Benefactors of a robust production system with zero loss. TPM breathes new life into tired TQ programmes.
Line operators	Time and willingness to learn.	New skills, a better working life and a new inter dependency with the company, such that with the new TPM skills they are attracted to other businesses if they are not cared for by the current employer.
Production engineers	Equipment and process designs as well as technical knowledge.	Better process designs and an emphasis on proactive production system development rather than reactive planning.
Maintainers & engineers	Skills, time and enthusiasm.	A more routinised loading of effort but with a much greater proportion of time spent using diagnostic skills to solve problems and contribute to the competitive success of the firm.
Financial controllers	Costing of improvements and savings made.	A stabilised and standardised process upon which to build budgets and attract new business.
Human resource specialists	Knowledge of employment contracts. Counselling of workers during the TPM process.	A more fluid and flexible contract of employment which is biased towards skills development, learning and greater willingness of workers to change.
Trade union officials	Technical skills and a new means of decision making and management within the factory.	Promotion of TPM as a means of up-skilling workers, securing the future of the factory and engaging in management–union productivity partnerships in a meaningful and measured way.
Asset suppliers	Specifications, latest advances and new processes to support the customer.	To eliminate all the waste associated with traditional machinery by learning from the past and building in new features that make the machine easier to work and more reliable. Writing of simple [Single Point Lessons] asset maintenance routines before machine is installed.

The OEE measure is however often abused and the subject of many discussions. Many managers tailor it, water it down, or try to inflate the figures. This is often the case at the beginning of the OEE launch when everything seems complex and too difficult. You should avoid this – no matter how bad the initial calculation. Pure calculations are needed to develop a trend in performance and to highlight target areas for projects which will readily ratchet up performance closer to a respectable 65% level and then, through sustained effort, to the 'world class' level. We will return to the centrality of this measure a bit later.

Box 8.2 Calculating OEE

The OEE Measure

The final OEE measure is derived from multiplying three percentages together. These are Availability % of the production line, its Quality Outturn % and its Speed Performance %.

$$\text{Availability \%} = \frac{(\text{Loading Time} - \text{Unplanned Downtime})}{\text{Loading Time}} {}^{*100\%}$$

$$\text{Quality Outturn \%} = \frac{(\text{Production Output} - \text{Defects})}{\text{Production Output}} {}^{*100\%}$$

$$\text{Speed Performance \%} = \frac{(\text{Cycle Time} * \text{pieces produced})}{\text{Operating Time}} {}^{*100\%}$$

It should be noted that Operating Time is the equivalent of (Loading Time – Unplanned Downtime), and, to illustrate how the final OEE is calculated, let us take a process which has an Availability of 95% with Quality of 92% and a Speed performance of 87%. So 95% times 92% times 87% yields an Overall Equipment Effectiveness of 76% – Not bad! But not 'world class'.

TPM pillar approach

There are eight major pillars which make up the basic TPM system and even though these have been extended in recent years they remain the fundamental building blocks and logic of the approach. These pillars are:

1 Safety and workplace discipline
2 Education in TPM
3 Overall equipment effectiveness improvement
4 Planned maintenance
5 Autonomous maintenance

6 Quality maintenance and methods control
7 TPM in the office and support function integration
8 Early equipment management

The eight pillars commence with the review and development of the safety pillar, from here most companies will engage in training and development of both the management and workers in TPM (TPM education), before measuring overall equipment effectiveness and getting to grips with what maintenance should be conducted and by whom (planned maintenance and autonomous maintenance). Neither of these pillars will be complete, nor go beyond the initial pilot programme before each of the others commences or those involved are forced to confront the development of the quality maintenance pillar (to improve equipment effectiveness through the elimination of the sources of defect production) and the stabilisation/standardisation of processes and documentation (TPM in the office). After a period of these pillars working effectively, then the outputs can be distilled into the policies and process which governs early equipment management. From this point on, the interplay between the pillars is so high that the overall demarcation of pillars becomes increasingly blurred. It is at this point that employee skills are so good that in many ways it does not truly matter from which pillar the improvement activity was born nor which benefits – it is at this point that TPM becomes a fully functioning system.

Each pillar reduces further the waste in the factory when implemented and at a point of functionality – let alone a perfect pillar implementation – performance will improve enormously. Some of these gains will derive from morale uplifts which if left will soon decay but most will result from the application of new skills in a systematic, quality assured and safe manner.

Pausing for thought

Before we go any further, let us put into context what adopting the pillar system means in a language that can be understood by all. If, until this point, you have defined the value of operators as individuals who make good products and continuously improve their activities to increase value, then you would probably agree that the definition of value for the role of the technician and indirect operations management staff is derived through use of their diagnostic skills. This is a change process that is often forgotten especially by 'fashion-led' companies who will 'do' TPM because it is popular but will not exploit its full value as they do not really understand what it is and how best to use it. The TPM process redefines the 'technician' and 'indirect staff' who support operations by creating the environment in which the specialist knowledge and skills of the technician can be used effectively. This is a radical change and for many firms this transition process has to be managed properly. It is here (in the technical and office 'staff') that you will find the resistance.

Table 8.3 *Eight major pillars*

Pillar	Purpose	Practices & benefits
Safety and workplace organisation pillar	To review workplace organisation, working practices and the causes of accidents to develop an effective control and problem-solving system.	Workplace organisation. Cause and effect analysis. Employee training. Visual management. Discipline for persistent offenders. Control of 'isolation' and 'lock off' procedures.
Education and training	To collate and develop the learning library for the TPM initiative. To write and quality assure safety of the materials and all processes.	Single point lessons. Videos, training facilities and document control. Savings calculations.
Improvement of equipment effectiveness	To measure OEE performance, track trends and highlight improvement areas.	OEE reviews and visual measurement (trends). Reports to management.
Planned maintenance	To establish, control and review the assets and the planned schedules for technical staff (and autonomous activities).	Analysis of failure points. Calculation of mean time between failures and scheduling of asset care. Identification of machine improvements.
Autonomous maintenance	To create a quality assured and safe process for front-line operator team maintenance and improvement activities.	Clean and inspect routines and potential lubrication routines. Safety management. Visual management.
Quality maintenance	To analyse and eliminate the various sources of defects and equipment related problems resulting in quality losses.	Data analysis and small group work. Machine improvements and maintenance prevention.
TPM in the office	To develop a robust system of document and process control. To build an effective library of improvement activity and to streamline the administration of the system.	Control of documentation, procedures and office-based improvement programmes.
Early equipment management (EEM)	To introduce new technology in a controlled and expedient manner with zero losses and optimised performance on commissioning.	Current asset fault analysis. Cross-functional project management. Technology-advances scanning. Supplier involvement

Resistance typically comes from uncertainty, inability to articulate what effects TPM will have and also from employees (who deliberately act like 'dip sticks') who will use this initiative to start rumours of cutting back overtime hours and all manner of propaganda. Most of these rumours are wrong! It does not decrease work – in the short term it increases it. The only issue which is truly contentious concerns the payment of new skills acquired and how this affects the legal contract of employment. This is a management concern and well worth a joint management–union study committee which should be established to understand the implications of TPM decisions and the pillars. An approach termed the '5 Hows' is great and works just like the five 'Whys' we discussed earlier[1] except you ask 'How?' to get the activities needed and true discussion points understood by the study group so that recommendations can be made ahead of time and before the implementation process becomes slowed down awaiting a management policy to be formed.

Let us go back and imagine the stereotype of the Western technician. Horrible isn't it? You are probably thinking of an obstinate worker who spends their time under or in machines just getting them to run again without getting to heavily involved in eliminating the sources of failure. Add to this a generally scruffy appearance (the mandatory blue boiler suit) and a bit of an attitude when it comes to the intelligence of the average operator and you're probably quite close to the typical stereotype. What a waste! These individuals went to technical college and have developed a specialist skill concerning the functioning and operation of machinery. What the stereotype depicts is a person who does not use this unique skill to add value within the business, but has given up 'fighting the system' to accept the 'fix it and buzz off to the next crisis' approach. The art of TPM is getting these individuals engaged, motivated and willing to transfer some of their 'routine' skills, from which they derive no real pleasure, to operator teams in the form of standard procedures and for the technicians to engage in 'value adding'. It means the technical person must be involved and act as a means of safety and quality assurance in allowing their skills to be passed between grades. There is absolutely no point in passing on skills to be used by others if they will not use them safely or to the standards of quality that a technical person would expect. Value adding for technicians therefore involves project work and using their skill set to solve, eliminate and improve the sources of sporadic and long-term under-performance, as well as to be at the heart of the quality assurance of the TPM programme (especially skills and auditing). No TPM programmes, even the really well-planned ones work out exactly when it comes to the human issues of skills, pay and contracts. Some companies do not like operators conducting lubrication routines (and I agree that this needs a level of competence beyond the clean and inspect routines). As such, a clever manager will add in 'discussion time' and time to formulate policies and proposals at these sticky times – its fair and a good way of incorporating all the stakeholders needed to make the roll out of TPM a success.

Having said all this, there is a further dimension to TPM. You will have seen the many basic pillars and what they contain – the big learning point is this programme

[1] For each specific solution identified, i.e. to improve machine reliability, the question 'How?' is posed to achieve an answer then it is posed another four times to identify, in detail, exactly what countermeasures should be introduced.

will take considerable time to reach a basic functional system. About three years is the average, you will be told by a Japanese sensei (master engineer) to get a good functioning system and that is by concentrating exclusively on TPM. But do not get depressed at this implementation time, as the benefits, which each manager will be interested in, come a lot quicker and most sources of waste and loss will start to move in the right direction (towards 'zero') and that means 'money saved' much sooner.

Lean TPM implementation 'blue print' (Rich, 1999)

Many of the terms we use today to describe 'best practice' and 'world class' manufacturing include new initiatives such as lean or six sigma or TPM. These approaches are religiously defended by individuals as better than the others, but in truth they are all aspects of the same 'system' and each shares a common toolbox of methods – and you often get caught in discussions about which technique comes from which programme. It does not matter. If it is a commercial improvement tool then who really cares? – its use is of more value than who invented it. Thankfully, the three programmes cited share the same basic DNA and they cannot exist without each other.

The beginning of the TPM journey starts with the introduction of the CANDO system across the factory to commence the safety pillar and from here it is possible to create the series of analyses needed to establish the 'safety pillar'. These activities, involving representatives from each team, will assess the nature and incidence of accidents to reduce these to zero in the new disciplined and organised workplace. With the safety pillar in motion, it is possible, from the initial CANDO system to grow the problem-solving skills of the workforce using more formalised problem-solving activities to commence the quality pillar and to rid the factory of the low-hanging 'fruit' represented by the many simple problems that beset operations staff. At this point, it is also possible to begin the collection of OEE data to understand the sporadic and chronic losses of performance capability within the production system. The collection of this data will be of most use to the technical teams as they engage in setting planned maintenance routines and to extend the meantime between line failures. During both these stages, the technical personnel are involved with the operations teams to add a new dimension to problem-solving activities. The purpose of this involvement is to add to the quality of solutions on the passage to zero quality losses and to avoid a Western fixation with the quantity of solutions generated. The quality of solutions has a direct impact on the OEE calculations and therefore action and result can be seen by technical and operations staff. To hold these gains, the best way of standardising practices, is to document the changes and new procedures using the single point lesson (SPL) method.

With the early safety and OEE tracking pillars in place to assess waste levels in the factory, the next stage is to seriously engage in the TPM process by understanding the losses within the production process using the OEE measure. The extension of this pillar provides enormous insight into the production facility and its failings. It brings with it also an awareness of what assets need to be restored to a good working condition that is process capable. Many businesses fail to conduct this level of analysis and eagerly engage in 'autonomous maintenance', which simply routinises with procedures the daily team work around the asset – it misses out the need to restore certain parts or indeed the entire machine. Changing the emphasis of the cause and effect chart to focus not on the "M's" (materials, machines etc.) but those of availability, quality and speed losses of the line increases further the resolution and focus on the improvement teams and allows gains to be expressed as commercial benefits. The milestone here is to track the mean time between failures or the amount of time that the line runs without intervention. These gains are not enough to sustain 'zero intervention' at the line though and this capability comes from adding the planned maintenance and autonomous maintenance pillars.

The target of these purer TPM activities can be focused, in complex factories, by an initial critical analysis of the asset/production line, and as such can identify the machinery with the most important impact on the commercial trading of the firm. Here it is helpful to use a '2-by-2' matrix to understand which assets are both critical to quality (CTQ) and critical to customer service (CTCS). All too often, consultants head for 'non-bottlenecks' or 'non-workhorse' lines where they herald enormous savings from machines which do not influence customer service. This must be avoided. Lean TPM companies will look to assure the flow of materials through the value stream and therefore will work on perfecting the OEE performances at machines before and after the bottlenecks, whilst looking for the best way of improving the bottleneck itself and getting more out of the production system for the same level of inputs.

The planned maintenance pillar involves a series of asset condition assessments to identify what aspects of the equipment need to be restored from a worn condition to a good standard. These activities are supported by using a rating system at the asset level concerning reliability, ease of operations, maintainability and such like. Even though the assessment is subjective – it is a starting point. It can be reinforced with reference to OEE data that have been collected, but this has the weakness which results from data inaccuracies as operator teams identify the reasons which they think stopped production rather than a true knowledge of the cause (not the visual symptom). By engaging this pillar, even further weaknesses in the production system and its data collection processes will be exposed (including the lack of a robust asset register, poor data from the reactive works orders system, poor handling or critical asset spares and ineffective planning of statutory activities like portable electronic appliance testing PEAT). This is to be expected, most Western systems have operated for years with limited budgets and poor care taken over the design of maintenance support systems. It is an area of management that can easily be fixed and revisited in the same manner that value stream mapping (McCarthy and Rich, 2004) shows managers how products flow even though most managers have been in their jobs for years.

Box 8.3 Single-point lesson (SPL)

A very powerful yet often ignored feature of a good TPM system is the use of Single Point Lessons (SPLs) and these documents are very effective ways of increasing operator skills without the hassle of extensive 'lectures' in training rooms. It is a single piece of A4 paper which contains all the instructions needed to conduct a task (these documents have been quality assured and checked from a safety management perspective). They are used to substitute words for diagrams and pictures such that they are easy to follow and represent efficient processes. They are also the bedrock of teaching and costing improvements.

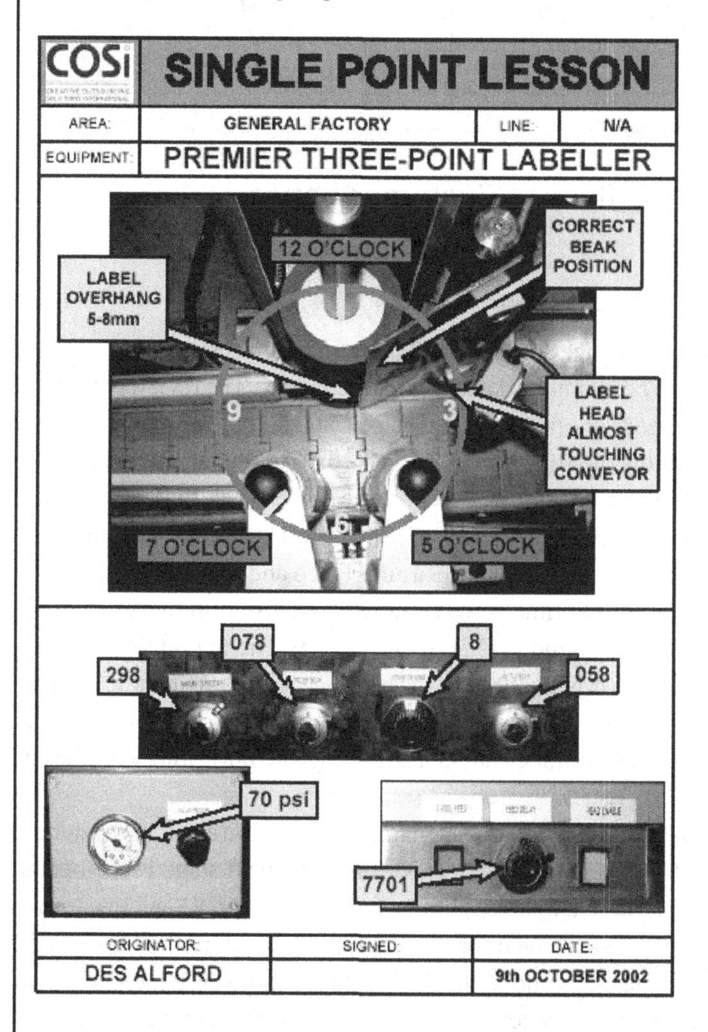

This one, the first edition which has been improved three times since, was supplied courtesy of the operations teams at Creative Outsourcing Solutions International (Des Alford, 2002).

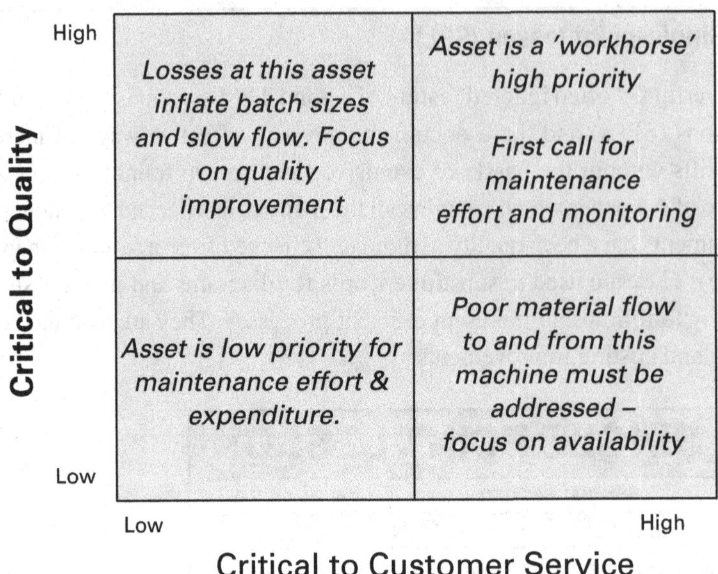

Figure 8.3 Asset criticality

The planned maintenance activities will therefore concentrate upon what aspects of the machine need to be maintained and at what frequency. The reviews of the assets will also highlight the need for operator teams to be trained in these aspects of the machine if only to identify how the sub-assets like 'gearboxes' function and what visible signs of wear they exhibit (oil leaks). As such, these systems will allow each asset to be assessed in terms of breakdown analysis (Sporadic losses), the identification of the weak parts of the machine, the countermeasures taken and the trends established, which all support the extension of time between failure and also improve productivity levels (witnessed by an upward trend in OEE performance). Having established the planned maintenance activities, it is possible to determine which tasks can be simplified and, through training and certification, can be passed across to competent operators. These activities tend to involve minor cleaning, inspecting and lubricating activities (the latter is often a contentious issue).

The development of the autonomous maintenance activities (those routines that are destined to be self-managed by line teams) forms the next stage in the operator team development. So far these employees have engaged in CANDO and low level problem solving and the next logical stage for them is to be trained in identifying the types of contamination that can be found in the factory (indications of impending asset failure and to use the works order system to warn of this event). With the correct safety and 'lock out' procedures followed, it is possible to train operators in signs of contamination (product spillage, water, oil etc.) and then to tour the production line identifying these sources before being trained and facilitated in asset cleaning. The latter requires a lot of maintenance effort to identify all the parts of the machinery that can be cleaned, the cleaning materials and means, and also the areas of the machine which have, even when isolated, safety issues (cutting blades, heat sources). It is also vital that, at this stage of preparation, the teams are instructed in the basic functionality of the machine so they become familiar with the elements of the machine and can later attribute failings to this level of the asset. These functional areas will include gearboxes, filling heads, sealing units, pumps, motors and all manner of major sub-asset types.

With the right materials and personal protective equipment (PPE), it is possible to remove the machine guards, identify a lot more contamination and then to clean the machine safely. Here it is important to take photographs during this stage of cleaning to show before/after illustrations, but, more importantly, to generate new single point lessons which show the correct standard of cleanliness, cleaning points and cleaning methods as determined by the maintainers facilitating the programme of learning. As such, the cleaning routines of the asset may be identified and the frequency of the need to clean established relatively quickly. After the exercise it is important to review how the sources of contamination can be eliminated (eliminating the need to clean) or contained to prevent contamination beyond the place where the source can be found. It is also important that the teams begin to identify the sources of contamination and their sub-asset classification (oil leaking from the seals of the gearbox or over-greasing of roller bearings leading to splatter and dripping of contamination on to the bed of the machine). The identification of contamination and cleaning routines is also supported by a review of the 'difficult to manage' parts of the machine such that these cleaning activities are targeted to eliminate the need to clean.

Most groups find this exercise stimulating and a natural extension of the problem-solving routines they have already learned (not 'just another initiative'). Teams will naturally propose machine improvements which will yield results, in their eyes, but this has to be 'sanity checked' by technical staff to ensure that these solutions are of good quality and safe for the machine and the person. A typical outcome of this cleaning activity is that more corrective actions will be identified through the nature of cleaning as an inspection activity (missing covers, loose machine parts, etc.), but also many teams will suggest ideas to replace machine covers with transparent ones (to see what is happening inside the machine) or to introduce a series of 'skirts' around the machine which act as barriers to prevent the spread of contamination. These improvements, where accepted, should be introduced reasonably quickly to maintain the enthusiasm of the team. The cleaning routines may then be reviewed again to see how much time has been eliminated and new standards put in place (also a number of works orders will be raised to improve certain aspects of the equipment and these must be dealt with by technical staff too).

Many 'purer' TPM companies will then instruct the operator teams, after a period of getting familiar with the cleaning routines, in the science of lubricating the asset. However, experience with operator teams would lead us to support the next team activity to be the identification and control of inspection devices. These devices come in all different shapes/sizes and they have been installed on the machines to control key sub-systems. Examples include regulator gauges and sight-glasses which need to be monitored. A common approach is taken to that of the cleaning routines and, under safe conditions, the teams look for the many different types of inspection devices in place. A note is taken of the position of each device and the standards of machine operation are then checked (air pressure regulator at the filling head needs to be within 12 to 14 PSI[2]). Through simple visual management techniques, these standards can be made easy to see (marking of dials or high–low levels on asset oil reservoirs). These simple additions cut inspection time down and when every device has been identified and made visual, it is possible to add these to the cleaning routines and to devise the most time-efficient circuit of clean and inspect activities around the machine.

We are now at a major sticking point. After a period of using the new clean and inspect standards, the next phase of deploying simple front line maintenance to the team includes lubrication. Again concerns must be raised at this point, as these activities are vital to asset performance (poor lubrication is the main source of machine failure) and cannot be neglected. Effective lubrication demands that all lubrication points on the machinery are identified, as too are the means of lubrication and the correct form of lubrication as well as the means of error-proofing the system such that only the right form and method are used by operations teams. Again visual management and the allocation of materials to the lines (with controls on the issuing of materials by engineering) must be engaged (colour coding points to show the different lubrication types, i.e. green paint near the point to show grease must be used, blue to show a neat synthetic oil and, say, orange to show where a light maintenance oil is to be used). The frequency of lubrication can also be determined by the technical staff and these too can be added to the existing clean and inspect routines to reach a point of autonomous checking. A note of caution concerns the quality assurance of the operator routines and the need for maintenance to periodically check to see the work is conducted correctly, but more importantly that the process has been mistake proofed to prevent the wrong type of lubricant being used or excessive amounts being consumed (this is added to the planned maintenance and audit task lists). The invigoration of OEE-based problem solving also reinforces any slippages in autonomous maintenance activities and any remaining factory contamination will be evident in CANDO scores. Furthermore, it is possible to revisit the single-point lessons covering cleaning, lubricating and inspecting activities to streamline them into a single document, which shows the optimal passage around a machine. Here it is also possible to engage in quality maintenance activities and to add to operator skills by looking for the parts of the machine which cause defects and to identify what needs to be changed to reduce to zero any scrap or rework levels. In this process of learning, the operator teams see no inconsistency as each new phase simply adds a new set of routines (with much greater attention to detail) than the last. It is therefore not shocking to the operators but a logical extension and new skill

[2] PSI denotes the calibration pounds per square inch of pressure.

set that is being given to them – which are totally at odds with the traditional 'quick fix' approach to improvements of the past.

Having conducted the training and having passed over front line responsibility for daily maintenance tasks, so that these can be conducted in less than five minutes per day per operator, a system of control is now in place with planned maintenance of a periodic kind conducted by specialists and front line operators dovetailing with them to provide constant attention to signs of abnormality. To support the planned and autonomous maintenance programmes, the office TPM pillar is established to write and control the procedures of the factory teams, to collate data and to provide a range of information routines (archiving asset information such as design drawings, costs, critical spares, single point lessons etc.) and to provide management with reports. This pillar can be started as soon as TPM begins at the safety stage, but really provides value as the planned and autonomous pillars evolve. This provides the capability that traditional businesses lack – information about the value stream and its performance. With an effective control system in place, it is now possible, with the data collected and improvements undertaken, to engage in early equipment management. The last pillar is by no means the easiest and is closely linked to office TPM – this pillar uses information about existing machine performance, weaknesses, improvements undertaken and standardisation of activities to eliminate all the potential sources of failure when specifying the next generation of machines. In this manner the company does not have to re-enter the problem-solving stage again (but can design out problems with suppliers) and commission assets in a much quicker time-scale. The latter is a major competitive capability and means the business will be 'quicker to market' with new products requiring new investments in technology. For many TPM companies, this involves a 'vertical start-up' such that assets are commissioned quickly and without problems so as to reach the 'steady state' production output in only a matter of hours after taking the machinery out of its box. This vastly reduces the pay back of equipment and increases business revenues, which breaks the final myth that maintenance is a cost centre and has little to do with earning a profit for the firm.

From what has been discussed it is possible to discern several stages of TPM development:

1 **From chaotic to stable state** of asset performance using corrective actions and introducing the basic planned and autonomous maintenance control procedures with machine improvements (elimination of known failure sources).
2 **From steady state to optimised state** through the identification of chronic machine failures (not sporadic breakdowns but the consistent underachievement of 100% OEE due to machine minor stops). To close the gap between the current performance and the ideal, quality maintenance, planned maintenance and autonomous maintenance must combine problem-solving activities to result in equipment upgrades and procedures which stop seconds being lost at the machine rather than minutes or hours.
3 **The extension of the asset life** through improving OEE performance and using data to identify the weakest elements of the machinery to result in upgrades and

standardisation of asset subsystems. At this point, the data systems for TPM are most effective and can be used to 'robust' the asset and specify the standards needed for the next generation.

In this manner, it is not possible for businesses to have 'done' TPM and to dismiss it as being achieved, nor is it possible to engage meaningfully in more advanced patterns of wear analysis and complex solutions when assets are still being controlled by, in a nice sense, 'ignorant operators'. As such, TPM is the hardest of all the lean practices and methodologies, it is the longest to incubate and yet the most effective of weapons in stabilising business costs, improving performance and contributing customer value. It is one of the few programmes with a direct, rather than tenuous, link to quality, delivery, costs and design performance, but it is not the easiest and even the pioneers have not 'done' it yet but they do achieve near-perfect OEE performances for large parts of the working year.

Summary

So it is impossible to operate a lean system over the medium and long term without TPM and workers will fail to make their full contribution to business improvement processes. Neither will the value stream mapping approach prompt the building of the support systems needed to make, in a 'total' approach, the most effective contribution of maintenance-operations effort. Only when the lean production approach is combined with an enterprise-wide system of TPM are the full benefits of stockless and high quality value stream management practices capable of commercial exploitation and 'pay back'. However, the establishment of planned, quality and autonomous maintenance, guided by the OEE calculation, is just the first stage in a process of achieving 'zero loss' and these form a 'basic' but robust lean system upon which to roll out the programme to other value streams and also to extend these to suppliers in the war to optimise material flows. TPM, is a victim of its own name and the term 'maintenance' is the method to unlock the full potential of a lean system and when combined with practices such as level scheduling allow operations to work almost indefinitely without interruption as the company works ever closer to the 'fifth key' and the state of perfection.

Signposts and recommended reading

There is no way a writer could do justice to a change programme like TPM in just a chapter. Generations of Japanese master engineers have dedicated their lives to the study and refinement of TPM and this chapter was just to whet your appetite for more. In addition to those cited, the books that offer good insights into TPM include Tajiri and Gotoh (1999), Wireman (1992), and Shirose (1996a, 1996b, 1997).

9 Sustainability

Nicola Bateman and Ann Esain

Introduction

At LERC[1] we frequently receive enquiries about sustainability[2] of lean programmes. Specifically managers want to know what can be done to maintain the improvements already made and continue the drive to improve. What we would like to reply is, 'Take these three simple steps and you will have guaranteed sustainability.' Sadly all our experience and research shows that different organisations and even different activities within the same organisation have varying problems with sustainability and so a more contingent, less prescriptive approach has to be adopted.

This chapter will look at what we mean by sustainability, and how we know when we achieve it. At LERC we have also conducted some research in this area with the Society of Motor Manufacturers and Traders' (SMMT) Industry Forum (IF) and we will be outlining the findings of this research and some additional findings from our Learn 2 companies (Bateman, 2001). The Learn 2 companies were aware of the IF findings and had the opportunity to implement them. We conducted some research to see how effective they had been and explore the opinions of the change agents at each of the companies. The outcome of this research was less definitive and highlighted some of the difficulties with applying a formal model to sustainability of improvements. The chapter concludes by exploring how the research findings can be used by managers in their own organisations.

Box 9.1 shows some definitions used in this chapter. In many organisations these terms are interchangeable, but for clarity we have defined them in a narrower sense.

Defining sustainability

There are a number of organisational levels associated with the implementation of lean: strategic, value stream and operational (figure 9.1). The highest level deals with the

[1] See also the EPSRC funded Innovative Manufacturing Research Centre at Cardiff University and its central theme of economic sustainability of manufacturing businesses.
[2] The concept of sustainability is based upon that used by Repenning and Sterman (2001).

Box 9.1 Definitions

Continuous improvement: The process of applying lean tools and techniques to continuously improve processes.

Process improvement (activity): A process improvement (PI) activity is a focused activity involving a small team usually 5–15 people who seek to apply lean tools and techniques over a short period, typically one to three months.

Kaizen: Japanese for continuous improvement but often applied in the phrase Kaizen activity to mean process improvement activity.

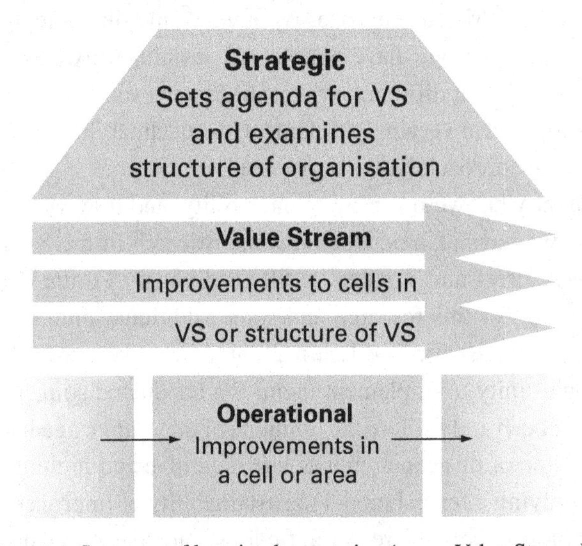

Figure 9.1 Structure of lean implementation (vs = Value Stream)

whole lean initiative and addresses the strategic and organisational issues and includes activities such as policy deployment and big picture mapping. At this level, with regard to sustainability, managers tend to struggle with keeping the impetus and not getting bogged down with day-to-day details. At this strategic level it is essentially a change management problem: what can be done to encourage people to accept the change programme and incorporate lean into their everyday lives?

Examining the base of the structure, operational lean implementation, means looking at what happens on the shopfloor and at similar operational functions. Specifically at LERC we have looked at sustainability of process improvement (PI) activities and the aspiration to evolve toward continuous improvement (CI) at cell level. This operational level is a good place to start examining sustainability as data can be gathered and groups of PI activities can be analysed.

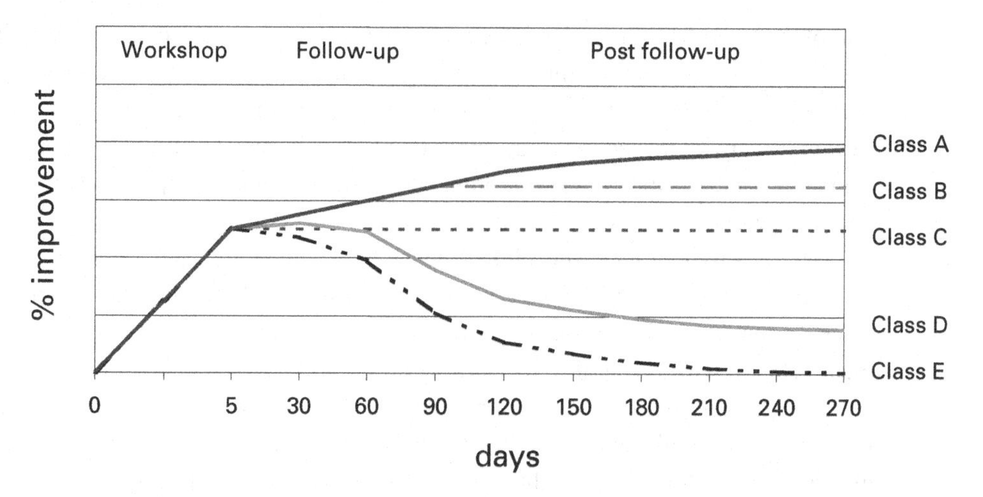

Figure 9.2 Sustainability of PI activities

The operational areas make up the value streams (VS) and improvements in individual areas should contribute to the performance of the total VS. In addition it will probably be necessary to look at the VS as a whole to simplify its structure, for example some elements may need removal or duplication to improve flow. Thus the VS level 'knits' together the operational and strategic levels by implementation of strategic plans and cross-company initiatives and by co-ordination of operational level improvements for a cohesive improvement approach.

Our starting point when we began thinking about what sustainability is, was to focus on PI activities and how sustainability of these activities can be achieved. It became clear that many different people had differing views on what sustainability is, so we had to define some categories of sustainability.

Examining what happens in a PI activity, the team usually do some analytical work data gathering and then they set objectives and do some planning. The next stage is to undertake a workshop. This is where the hands-on work takes place and where changes to the working practices of the area are implemented. At the end of the workshop – typically a period of two to five days – the area has changed its way of working and there is a 'to-do' list of jobs to be completed. The successful (sustaining) team manages to maintain the new way of working and close out the 'to do' list. The next stage is then to apply the lean tools and techniques as part of the everyday working practices of the area. If this is achieved, the team has gone from process improvement, a focused activity, to continuous improvement, whereby lean principles have become embedded in the everyday life of the team.

To show this progress, LERC constructed a model (figure 9.2) that indicates the range of attainment in sustainability from full continuous improvement to no sustainability, i.e. relapsing back to former processes. The model maps an activity over time from the start of the workshop, through the follow-up period and then extending over a period of a few months and indicates the level of improvement achieved. It is important to note when interpreting this model that there is no scale on the improvement axis. This is because we are interested in trends rather than actual levels of improvement, after all it is much easier to improve a poor company than a very good company where existing levels of performance are already high.

Class A activities have the highest level of sustainability and go on to develop continuous improvement from a process improvement activity. Essentially they have taken as much as they can from the activity and absorbed the lessons and applied them. This is why being able to learn as an organisation (see next chapter) is very important to sustainability as learning enables organisations to fully realise an activity and develop their own expertise. Class B activities have failed this learning hurdle, they have done well within the remit of the PI activity, but have not fully realised the potential of thinking beyond a bounded event. You could say they have treated the PI activity as a project with a finite end rather than a process which is on-going.

Class C activities are when the new way of working is maintained but the 'to-do' list is not closed out. They tend to occur where the team leader is strong in maintaining the new way of working, but there is a problem with resolving technical issues. This can occur for a number of reasons, including lack of technical support, pressure from senior managers not to put any further resources into the activity or very heavy demand to deliver customer orders, which means that there is no one free to be released to complete outstanding actions. Class D activities have the opposite strengths and weaknesses to Class C in that the 'to-do' list is closed out but the new way of working is not maintained. These type of activities occur where the role of the team or cell leader is not strong and so the team go back to the old ways of working. Often to succeed in closing out the actions, an individual single mindedly ploughs on despite the sustainability issues in the area.

In Class E activities the area rapidly goes back to the old ways of working and no substantial improvements can be detected. This does not mean all is lost, however, as often it is possible to revisit an area and have another go. In some organisations they are simply not ready to take on PI and having a go and failing takes them further along the path of knowing what to expect next time.

As an aside when assessing levels of sustainability it is very easy for change agents to focus on what could have been achieved (but failed to be) rather than how an area has actually improved. It is common to take this kind of negative approach and dwell on the could-have-beens. For assessing your own activities look at where you have come from rather than where you *could* have got to. After all, small steady improvements are what we are trying to achieve rather than revolutionary leaps. For organisations, the PI activity is often their first encounter with lean.

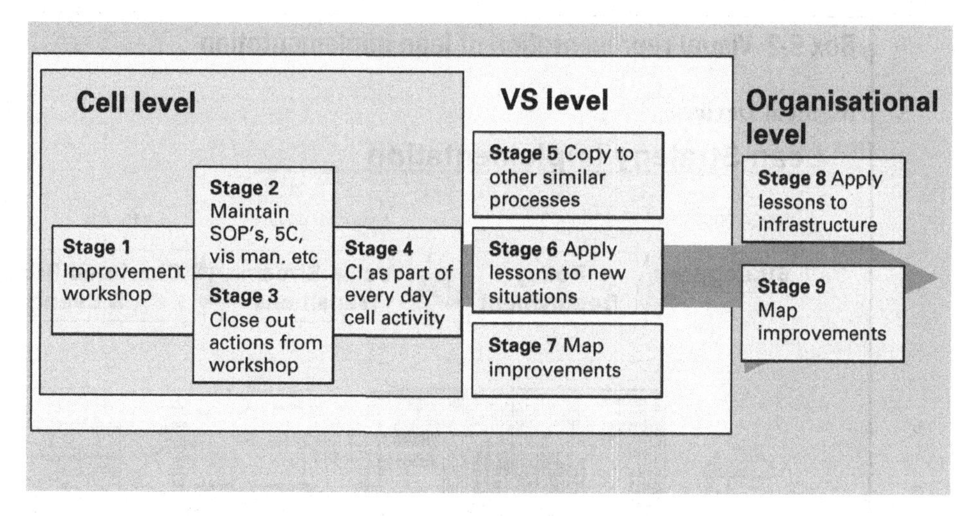

Figure 9.3 Rolling out from PI activities to strategic level

The next stage for many organisations is translating these types of activity into sustained improvement along the value stream and then across the organisation. This is the topic we shall explore next.

Sustainability, value streams and organisations

Translating successful PI activities into bottom line improvements along value streams requires consideration by management. Not only do you need to achieve successful improvements, but you need to do the right improvements in the right place and then tackle any additional issues that occur along the value stream. This is where mapping is useful and having management think about lean principles along the value stream is very important.

Figure 9.3 illustrates how to roll out the lean initiative from PI activities to the value stream. Stages 1 to 4 are described in the sustainability model (figure 9.2) and a class A activity equates to progressing to stage 4. Stage 5 standardises the process from the cell to other similar processes along the VS.

Stage 6 applies tools such as CANDO/5S or waste analysis to new situations, such as other cells in the VS, and also laterally across the VS to examine any infrastructural issues. The VS manager also needs to be co-ordinating these activities using a visible mapping tool and this is done in stage 7. An example of this has been used by Medical Devices who have linked their strategy through policy deployment, mapped their value streams and co-ordinated their operational improvement activities.

Box 9.2 Visual representation of lean implementation

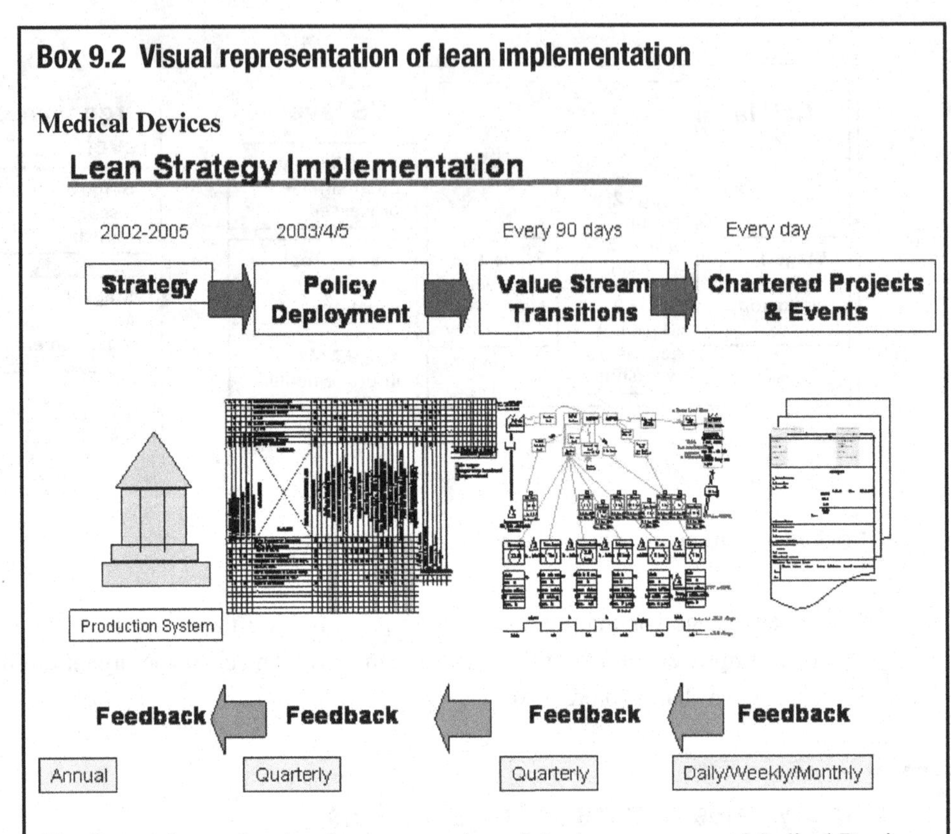

The figure shows the visual representation of the lean process at Medical Devices. The right-hand side of the picture shows improvement activities, these are mapped on to value stream maps. The value streams are shown in the policy deployment grid and this is determined by the strategy. There is feedback at each of these levels and reviews take place at appropriate intervals, as shown in the diagram

This bottom–up approach then needs to be transmitted to the organisational level (stage 8). At this stage it is difficult to be prescriptive, as organisations vary widely. What we are trying to understand is, what would a sustaining organisation look like, especially one building on its PI activities. So, having started to develop lean along the VS, the next stage is to take learning from the VS and translate it across the organisation. There will be substantial need to address organisational issues, such as training and technical support.

Essentially this is just following the PDCA loop making sure that approaches and 'best practise' are standardised. This type of bottom–up approach has to take place within a top–down strategic framework, described in chapter 3. Without the direction provided by policy deployment, it becomes difficult to sustain efforts in a coherent direction and staff become confused as to the point of their improvement activities.

- There should be a formal way of documenting ideas from the shop floor
- Ensure operators can make decisions in teams about the way they work

Contribution and buy-in

- Make sure that there is time dedicated to maintaining the CANDO standard every day
- Ensure there are measures to monitor the improvements made – at an appropriate level
- Managers (cell leaders and their managers) should stay focused on PI activity

Maintenance and focus

Figure 9.4 Enablers for Class A and B activities

LERC sustainability research

This research was conducted on behalf of the Society of Motor Manufacturers and Traders' (SMMT) Industry Forum (IF). Its purpose was to identify the common enablers associated with PI activity that attain high levels of sustainability. SMMT is the trade association for the automotive industry in the UK and is a non-profit making organisation working to improve the competitiveness of the automotive industry.

The aim of the research was to provide guidance to the IF engineers and CA (change agents) within the automotive sector companies. As such the research focused on enablers that could be addressed by people at these levels rather than exploring cultural issues that would be beyond their remit. The research was divided into two parts: quantitative and qualitative. The research surveyed 40 activities and the quantitative element identified those enablers that were associated with the activities that had higher levels of sustainability. The qualitative research questioned team leaders and change agents on what they felt were the inhibitors and enablers for sustainability.

The quantitative research initially compared those activities that obtained class A or B levels of sustainability (as defined in figure 9.2) with activities that obtained class C, D or E. These are shown in figure 9.4 These enablers can be divided into two categories: processes for ensuring 'contribution and buy-in' by the cell members and processes for maintaining the standard and continuing focus on the PI activity.

The processes for 'contribution and buy-in' are to ensure better decisions are made, based on 'hands on' knowledge of the processes, about the running of the cell. Increased ownership can be achieved as cell members have contributed to the decision-making process.

A process that formally documents ideas from the shop floor means dealing with issues raised by cell operators in a prescribed, thorough way. These issues would typically be ideas on how to improve the process, highlighting potential quality or

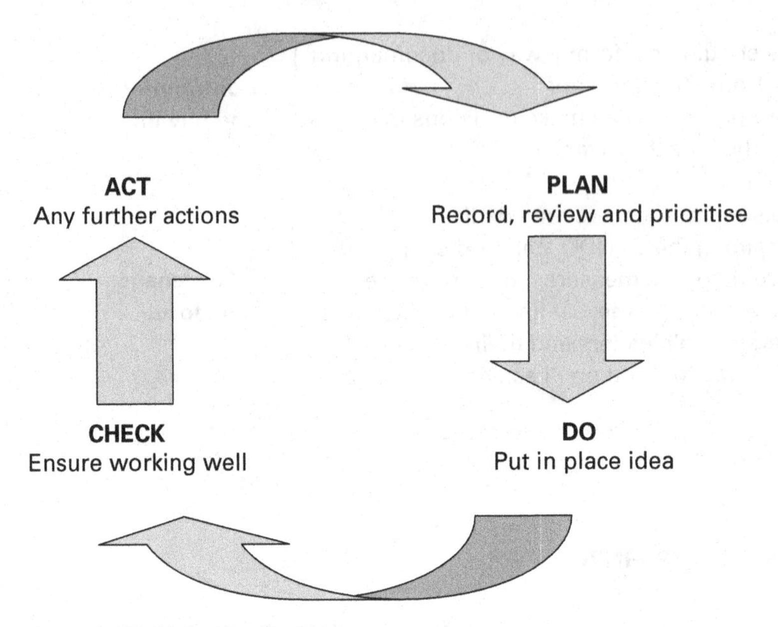

Figure 9.5 PDCA for shopfloor ideas

safety issues, or solving a current problem. The process for documentation has to be formal, but need not be onerous; examples include a flip chart in the cell area where operators can note down ideas or a review process at the end of every week where all operators can get direct access to a minuted meeting. Often it is not feasible for all operators to attend every week, particularly if there is a shift system in operation, but all operators should rotate attendance.

The important elements of this process map to a PDCA cycle and are: a method of recording ideas or issues, a method for reviewing the ideas and identifying those that are most important, acting upon the idea and closing it out and reviewing the action and raising any further actions required.

Simply relying on the cell leader to do something about operator suggestions is not sufficient because it is an informal process that depends on the cell leader being dedicated to taking notice of operator input. If the cell leader is under additional pressures or is replaced, the system for operator input may not be sufficiently robust to continue.

The second element of 'contribution and buy-in' consists of ensuring operators can take decisions in teams about their work place, this is important, as it allows the operators to have ownership of their area. This can typically be managed using a five to ten minute team brief at the beginning (or end) of each shift.

The three enablers associated with 'maintenance and focus' ensure that continuing effort is directed towards keeping the standard achieved in the improvement activity. Maintaining the standard achieved in the workshop is an essential part of CANDO – the first three elements the 'CAN' are dealt with in the workshop phase of the activity. The final two elements 'DO' – 'Discipline' and 'On going improvement' – can only be adhered to on a day-to-day basis. Discipline ensures that everyone adheres to this standard and typically CANDO check lists need to be developed as a reference standard. On-going improvement means that everyone participates with the processes of improving the CANDO standard. This diligence of approach is exemplified by one company change agent who commented, *'CANDO is part of life here'*.

Measurement of improvement achieved is also an important part of sustaining improvements made. It is a way of indicating what is important in an operational area and motivates people to support the measures. If no measure exists, people in the cell will not know what the standard is, and, as other priorities occur, the performance achieved in the workshop will gradually degrade. Measuring the performance can also highlight if any problems occur, such as untrained personnel, and, therefore, a solution can be developed. Care should be taken to ensure that measures are at the correct level, because the purpose of having measures is to allow people to take corrective action. If a measure aggregates the performance of several machines, this is useful for monitoring, but cannot directly tell an operator which machine has a problem.

'Managers staying focused on PI activities' is an important enabler because it reinforces the view that PI activity is still significant to the company, despite the initial workshop being completed. This enabler is directed particularly at the immediate managers to the cell, i.e. the cell leader and his or her manager. What this means is the managers do not get distracted into neglecting PI activities. This can occur with 'flavour of the month' syndrome when managers undertake one initiative, then undertake another without fully realising the first. There will always be situations when an emergency occurs and managers' attention is drawn elsewhere, for example a problem with raw materials or installing new machinery, but PI should always be returned to. This is exemplified by one company in the research group that stated, *'This [PI] is something we have to keep coming back to.'*

To further develop an area so that it goes on to Continuous Improvement (shown as class A in figure 9.2), enablers in addition to those shown in figure 9.4 are required. These enablers were identified by comparing the enablers associated with class A activities with those associated with class B to E activities. The enablers identified are shown in figure 9.6.

These enablers can be summarised into three categories; 'consistency and buy-in', 'strategy' and 'factory-level support and focus'. The first of these, 'consistency and buy-in', should be achieved by introducing changes to all cell members using a formal method. This is necessary because it is frequently not possible for all people who work in an area to be involved in all process improvement activities. The type of formal method required is a short presentation, usually by the shift representative in the PI activity, to each shift to outline the new features and working practices of the cell. This type of presentation can be supported by documentation, such as written standard operations and practical demonstration.

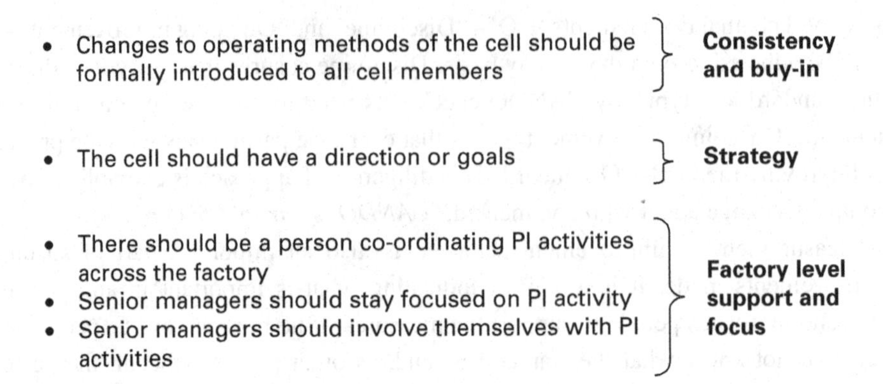

Figure 9.6 Enablers for Class A activities

'Strategy' for class A activities is particularly important because it guides the further activities that are an essential part of continuous improvement and ensures consistency in these improvements. To achieve this, it is important that the PI activity in the cell has direction. This would normally take the form of a statement with numerical goals derived from the manufacturing strategy, however not all organisations have a formal strategy development process and so a more informally derived direction is sufficient. Improvement teams need direction so that they can identify what issues are important for the company. Direction, in addition to internal issues of the cell, provides the improvement team with future actions.

For on-going improvement, additional 'factory-level support and focus' is required. This should take the form of a person co-ordinating PI activities. The co-ordination role was defined as requiring at least 30 per cent of a co-ordinators time. This person should not be doing the improvements, but rather facilitating and aiding cell leaders and improvement teams. In addition, senior managers, for example the manufacturing and managing directors, should be involved in PI activities by attending improvement activity meetings and by looking at how process improvement and lean principles can be integrated into the organisation of the company. Maintaining focus should be achieved by reviewing the progress of the shop floor as a whole, not imposing unnecessary additional initiatives, and as much as possible by protecting improvement activities from external interference. Focus can also be maintained by having the improvements made as part of the performance measures for the factory.

The second part of this research was qualitative and asked change agents and cell leaders what they thought were the inhibitors and enablers for sustainability. The general theme that emerged from this part of the research was that nearly all the companies reported that having enough resources – having the right people to run an activity – was an issue. What was interesting was the more successful companies addressed this issue by taking positive steps to resolve it. Managers also tended to be able to describe their inhibitors more precisely than their enablers; typically managers would describe as inhibitors *'not enough technical support'*, whereas enablers would be described more generally as *'open minded culture'*. This appears to indicate that managers are all too well aware what inhibits activities, but find it hard to understand what enables sustainability.

Learn 2

To develop our understanding of sustainability and assist the Learn 2 companies in examining their sustainability issues, each of the participating companies were surveyed and asked about the sustainability of their activities. The team leader for each area and the change agent for each company were asked what they felt were the significant inhibitors and enablers and were there any issues with engaging individuals within the teams. Their comments have been grouped into themes to indicate areas which can affect sustainability, the enablers are shown in figure 9.7 and the inhibitors are shown in figure 9.8. We acknowledge that this will provide a higher emphasis for those companies that conducted more activities and involved more people, but it was felt that we needed to capture as much feedback as possible to provide a good overall picture. Comments from team leaders and change agents are quoted here and we hope will provide a picture of the typical issues associated with sustaining PI activities.

Enablers and inhibitors of sustainability

The most commonly cited enabler, in common with the IF research was resources (figure 9.7) Typical explanations were 'time given for (team) event' and 'resources allocated' and what respondents meant was the team who worked in the area that participated in the improvement activity were given the time to do the activity and follow-up properly. Additionally some respondents cited having a dedicated resource for PI/CI such as a full-time change agent as an enabler. Correspondingly lack of resources and lack of understanding by mangers of what resources are required were cited as the largest inhibitor. A typical solvable problem was that the PI activity was regarded as a training programme, whereby attendees could omit the odd half day and catch up later. Of course, this does not fit well with the team element at the heart of PI and CI. Also concerned with resources were issues with specific resources that were required to complete some technical changes. This was particularly raised by team leaders who cited, *'engineering input slow and staggered'* and *'poor engineering support'*. This is not surprising as most PI activities tend to raise a whole raft of engineering issues, which creates a peak in demand for the engineering or maintenance function. Organisations that do not plan for such as rise in demand are bound to encounter problems with meeting requests for resources.

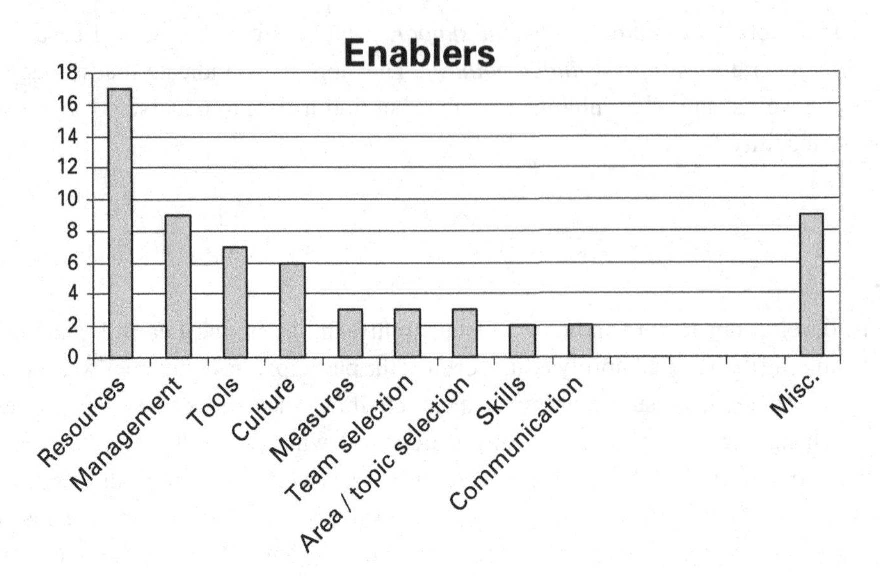

Figure 9.7 Frequency graph of enablers

A similar topic that is linked into resources is the 'volume versus improvement' inhibitor. This was not cited that frequently, see figure 9.8, but underlies much of the behaviour associated with resource issues. Essentially people in operations departments are principally concerned with delivery, i.e. what they are making today. So there is always a conflict between meeting today's targets and planning for tomorrow. Typical quotes are, *'Push on production takes up people's time'* and *'Focus on delivery'*. This is a perennial problem and the reason why team leaders, who are all too aware of day-to-day production pressures, can be reluctant to start PI in the first place.

The next most frequently cited enabler was management support and buy-in. This was a category that reflected similar feelings on the part of the change agents and team leaders that managers had supported the programme. Cited quotes include, *'top level management buy-in'* and *'support by senior management'*. Interestingly one respondent stated rather damningly *'top level pet projects enable change – sporadic focus on certain projects enables sustainability'*, implying that support was the whim of top level managers rather than part of a strategic plan. Looking at the corresponding inhibitors indicates what happens when top management support is not felt: *'lack of interest from top level managers'* and *'loss of focus at senior level'*. Generally the inhibitors implied that, whereas top managers may have initially supported lean activities, they found the long-term commitment required difficult to maintain. Also in the inhibitors section there were a few comments on managers' lack of knowledge of specific lean tools which meant that they were unable to fully understand the problems associated with lean implementation.

Figure 9.8 Frequency graph for inhibitors

Tools were the next group of enablers to be identified and these centred around CANDO check sheets and audits, TPM 'T' cards and a method to follow up suggestions. Comments focused on systemising lean tools, i.e. making them part of the ordinary life of the focus area. There was no equivalent inhibitor cited.

Culture was frequently cited both as an inhibitor and enabler, with team leaders and change agents either citing an enabling culture, such as *'Company culture of change and accessibility of change'* or an inhibiting culture *'kaizen is not really shopfloor culture'*. Attitude was identified as a separate inhibitor because negativity and cynicism from individuals was specifically mentioned, although this could just be an extension of the company culture. It is interesting to note that different people from the same company or even on the same activity would contradict each other with regard to culture – with one person identifying an enabling culture and another an inhibiting culture.

Measurement was cited as an enabler as it allowed people to be *'sold projects'* and so scarce resources could be focused on the correct areas. The corresponding inhibitors identified the need for good measurement and also highlighted the need for operational measures to be linked to strategic planning and thus avoid *'scope creep'*. Related to measures in general, the accounting system of one company in particular was identified as an inhibitor. In this case the way that operator hours were accounted for was a problem, in that it specifically discouraged managers from allowing their operators to do CI. This example highlights the importance of having measures that drive the right sort of behaviours.

Area/topic selection was identified both as an inhibitor and an enabler, essentially the message here is to choose the right area and focus for improvement. Quotes include, *'size of area – make sure not too large'* and *'pick a topic that will improve things for the operator'*. It was also difficult to motivate a team if they knew the area was due to be reorganised in any way in the near future, *'Significant changes in the (future) layout of the area meant that it was difficult to motivate people'*. Along similar lines, team selection was identified as an enabler, with *'small focused groups'* cited as well as selecting the team so that they would own the area after the activity.

Skills and lack of them were identified as inhibitors, particularly, understanding by first line managers. Some team leaders also thought that some of the lean tools were too difficult to be grasped by their operators. To some extent this was countered by other team leaders saying the opposite and praising the out-of-work skills brought into play to implement particularly CANDO activities.

Communication is often cited as a key part of change programmes, but it was not cited particularly frequently either as an enabler or as an inhibitor. A useful observation that is often true in lean is that teams *'forget how much we have done'* – often it is too easy to focus on what has not been achieved, rather than on what has been achieved.

There were considerably more inhibitors cited than enablers and a wider range of issues were raised. Unsurprisingly many of these were associated with fear of redundancies and defensiveness on the part of managers that their authority will be undermined. Regulation was also an issue for companies and it was a coincidence that two of the Learn 2 companies were both in different highly regulated industries. This proved a problem for both companies, *'Hugely regulated business makes change a painful process'* and not even the regulations themselves but the perception of them, *'perception of regulation requirements'*, caused problems.

The miscellaneous observations of the interviewed group were issues unique to the employing company. This group included *'four shift system'*, *'dust'* and *'large teams in cells'* for inhibitors and for enablers *'close links with unions'*, *'all staff involved'* and *'open forum'*.

Engaging people

In addition to asking about inhibitors and enablers to sustaining PI activities we were also interested in what happened at an individual level. So we asked team leaders and change agents to comment on issues associated with engaging individuals with the improvement process. In addition we also asked change agents what they had done to resolve any issues.

By far the biggest issue (figure 9.9) in engaging was negativity – and to gain some resolution of the problem we divided this into 'before activity' and 'during and after' activity. There was considerably more negativity before the activity and it is pleasing to see that this resolved to some extent by actually doing the activity. The before activity negativity generally stemmed from people's experience of previous change programmes

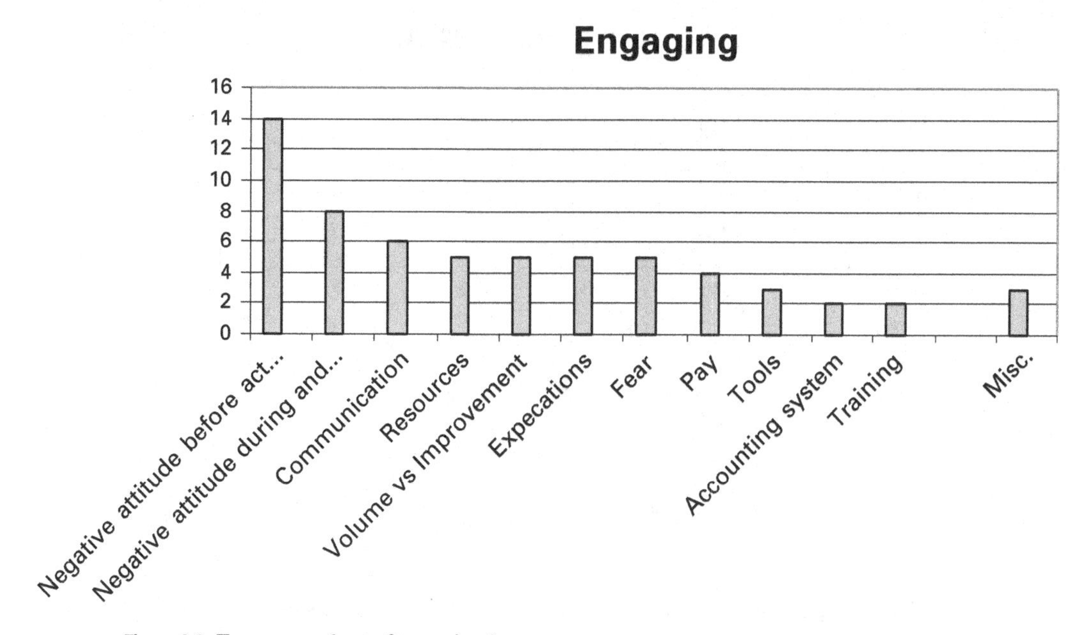

Figure 9.9 Frequency chart of engaging issues

with comments such as *'been here before'* and *'not another waste of time'*. This reflects previous attempts by managers to improve things, but somehow these attempts failed to succeed and hence a degree of cynicism crept in.

The degree of negativity 'during and after' the activity did seem to reduce from the 'before' category, with fewer comments in this column. It was not possible to eliminate cynicism altogether, as it appears to be embedded in some people's approach to work. The cause of this could simply be a cynical type of person or could be caused by working in an environment that promotes a cynical approach.

The communication, resources, fear and volume versus improvement categories simply reflect individual experiences of the general inhibitor and enabler issues previously identified. The volumes versus improvement inhibitor was exemplified by a change agent who commented, *'One area TL and supervisor would immediately dash off for a production concern no matter how trivial.'* This comment clearly demonstrates that, for this team leader and their supervisor, production niggles far outweighed in importance the need to improve, to the extent that the change agent felt that they were using it as an excuse not to address underlying problems in their area. The expectations group was mainly concerned with managing people's expectations of what was required of them in terms of commitment and in terms of what was achievable. *'The hardest part was maintaining their expectations after the initial period'* reflects a manager's problem of coping with raised expectations of the team once they had started the activity.

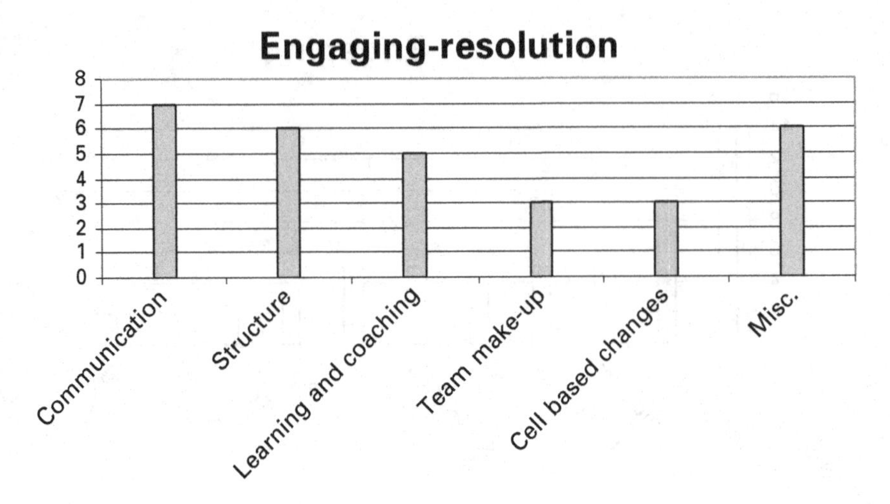

Figure 9.10 Frequency chart for resolution of engaging issues

Pay was a specific issue in one company, where participants expected additional pay for their lean activities. This is frequently raised in companies that have no previous history of operator participation, as operators perceive that broadening the scope of their responsibilities should be associated with a commensurate pay rise.

The lean tools CANDO and TPM were specifically raised as issues, with CANDO cited as *'dramatically improving the willingness to be involved'*, but TPM being viewed *'as not part of the programme'*.

The time taken to do training was cited as a problem and some people frequently found it frustrating where there was a time lag between training and the opportunity to apply it. The accounting system again reflected the comments made on the inhibitors and enablers.

Finally, the miscellaneous section addresses a small number of specific issues, but also includes one general and insightful comment, *'Business improvement engineers (change agents) get worn out – the task is very difficult'*, and this reflects observations made by researchers, namely that the role of the change agent, although exciting and challenging, can be a thankless task. If things go well, the team gets the praise; if things go badly, the change agent takes responsibility.

We also asked the change agent what they had done to resolve some of the issues with engaging individuals. The responses indicated that the change agent largely understood the problems, but had limited tools with which to solve them. The replies were categorised and are shown in a frequency chart in figure 9.10.

The most frequent positive action change agents could take was to improve the communication process. The comments in this group tended to fall into two categories, either changing what was communicated, particularly emphasising the benefits and proven processes, *'We had to prove the benefits to the team leader who is now more trusting'*, or changing the way things were communicated, for example showing how activities contributed to the whole strategy or explaining benefits in a more appropriate way, for example, *'Explain benefits in a way they could understand, highlight how this was a different approach'*.

Changes to structure of the PI activity (called structure in figure 9.10) were generally focused on fine tuning the PI approach and infrastructure. The most common comment was the need to appoint a full time CA to co-ordinate activities and communication. This is shown in *'Full time co-ordinators tackling communication across shifts and especially feedback, this resulted in a dramatic rise in self-ownership.'*

Learning and coaching was the next category, with most emphasis on the need to learn from activities and embed this in the PI approach, *'Constantly learning from mistakes made in PI projects'*. The coaching element generally focused on helping team leaders through the process improvement activity to fulfil their role as a team leader in a continuous improvement environment rather than in a traditional environment. This involved some teaching of new skills, but also illustrating the types of activities team leaders needed to do to encourage a CI environment. As stated *'Supervisors trained in business improvement to help sustainability and gain involvement of operators'*.

The next category 'team make-up' was concerned with getting the right mix of people on an activity. This generally reflects the need to have a team with the right types of people – enthusiastic and creative – but also to have people from the right departments to incorporate different shifts and technical people as required – *'Trying to form better, more rounded teams (cross functional) to better deal with issues'*.

The cell-based changes tended to be focused on putting the emphasis on local decision making, *'We left it to the local supervisors to set time for activity'* or spreading cell changes across the shift, *'Move from cell based to shift base thinking to increase people involvement'*.

The miscellaneous category included payment systems, trade unions, accounting systems and technical issues.

Resources

As part of this chapter we wanted to be able to provide managers with information to help them deal with some of the inhibitors identified above. Resources or lack of them are consistently identified as an issue for CI managers especially to fully roll-out an improvement activity. So we have included this section to give managers an idea as to what are the likely resources required (at shop floor level). We have aggregated a model, based on our experience, of how long this takes and the level of resources required.

The model was based on five areas, each having eight cells and each cell achieving A class sustainability and maintaining a 1% improvement per month. Areas start a PI activity, then launch their next one two months later; each subsequent activity kicks off

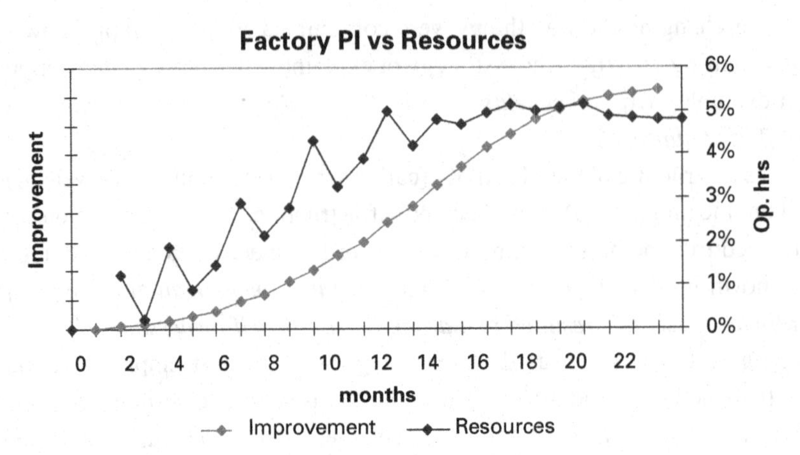

Figure 9.11 Resources required for full roll out of PI activities

at one month intervals. The first area starts PI activity in month 1, area 2 starts in month 3, and so on. The data are based on eight people in each cell and a 37.5 hour week.

Figure 9.11 shows the improvement that can be made after 24 months. At this stage the process reaches a steady state level of improvement of about 1% per month. It can also be seen that to reach this state requires 4.8% of shop-floor time. A 1% monthly improvement results in an annual improvement of 11.6% overall. To maintain this level of improvement is quite demanding and you can see the peaks in demand of resources on the graph as each area kicks off in months 1, 3, 6, 9 and 12, and all this does not take into account management time. So, before embarking on a PI process, really consider resources. If resources are not available, then the management can be perceived as not supporting lean.

Applying sustainability in your company

Drawing together the findings from our research and experience can be challenging, particularly in a subject that is highly contingent on the conditions in a company. So what we have tried to present here are some simple guidelines that managers can follow. They will require some adaptation and interpretation according to your own company's needs.

Processes for sustainability

This section interprets the findings of the SMMT IF research and the Learn 2 research. The four processes shown in table 9.1 are designed to encompass the enablers identified in the SMMT IF research and to 'hardwire' them into the processes that take place in

Table 9.1 *Processes to incorporate enablers*

Processes	At Least Class B					Class A				
	Record ideas formally	Cell managers stay focused	Time for CANDO	Team decision	Cell measurements	Formal Intro to new methods	Cell Aim	Person co-ordinating	Management involvement	Management focus
Role of cell leaders	✓	✓	✓	✓	✓	✓				
Team meetings			✓	✓	✓	✓				
Integration of PI, cascade measures and mapping		✓			✓		✓		✓	✓
Person co-ordinating							✓	✓	✓	✓

your operational areas. In addition we have tried to incorporate the feedback from the Learn 2 research where possible.

The first process is the role of the team leader, this should be to facilitate the change process and embody an atmosphere of continuous improvement. So specifically, the team leader should have a process whereby ideas from the shop floor are gathered, reviewed and fed back, thus following a PDCA loop. The team leader should ensure that the structure of the day includes a CANDO check list time and a team meeting in front of the communications board to review the day, discuss improvements and solutions. The team leader should ensure that the correct measures are included on the communications board and actions raised are closed out again following the PDCA loop. There will also be a need to address resource issues that are outside the team leader's operational area and the team leader's manager will probably need to be involved in this process. Finally the team leader will need to manage the process of introducing new working practices, such that everyone in the operational area understands them and can work to them.

The next process is having a daily team meeting. As mentioned above it is the team leader's role to lead them and ensure that they are well structured. A typical team meeting should not be too long – no more than 15 minutes – although there will be occasions when a more lengthy discussion is required, but the general rule should be to stick to a routine that does not draw in irrelevant discussion. To achieve this, it is useful to have a routine that addresses the relevant areas. Typically this might be:

- Yesterday's production – how did it go?
- Are there any problems to resolve?
- What do we need to do today?
- Recently implemented solutions – how are they working?
- Are there any new areas for improvement?
- Are there any issues regarding training, safety and holiday cover?

This can be done following the structure of the communications board, incorporating the measures displayed as appropriate.

The next process is to embed the activities of the cell into the whole lean plan. This means first giving it an aim and this aim should be related to its value stream. A typical aim might be to reduce takt time[3] by ten minutes. This might be achieved in a number of ways, but this would be decided by the team and their leader. A visual representation of the value stream should be displayed so team members can understand the role of their cell within the whole of the VS.

The final process is to have a person co-ordinating improvement activities. The role of this person is to facilitate, *not* to do the improvements. Typical facilitating roles include coaching team leaders, running improvement activities where required, standardising

[3] The average demand rate for products sold to customers which is set at the replenishment rate for manufacturing operations to keep pace with the rate of sales. See Monden (1998).

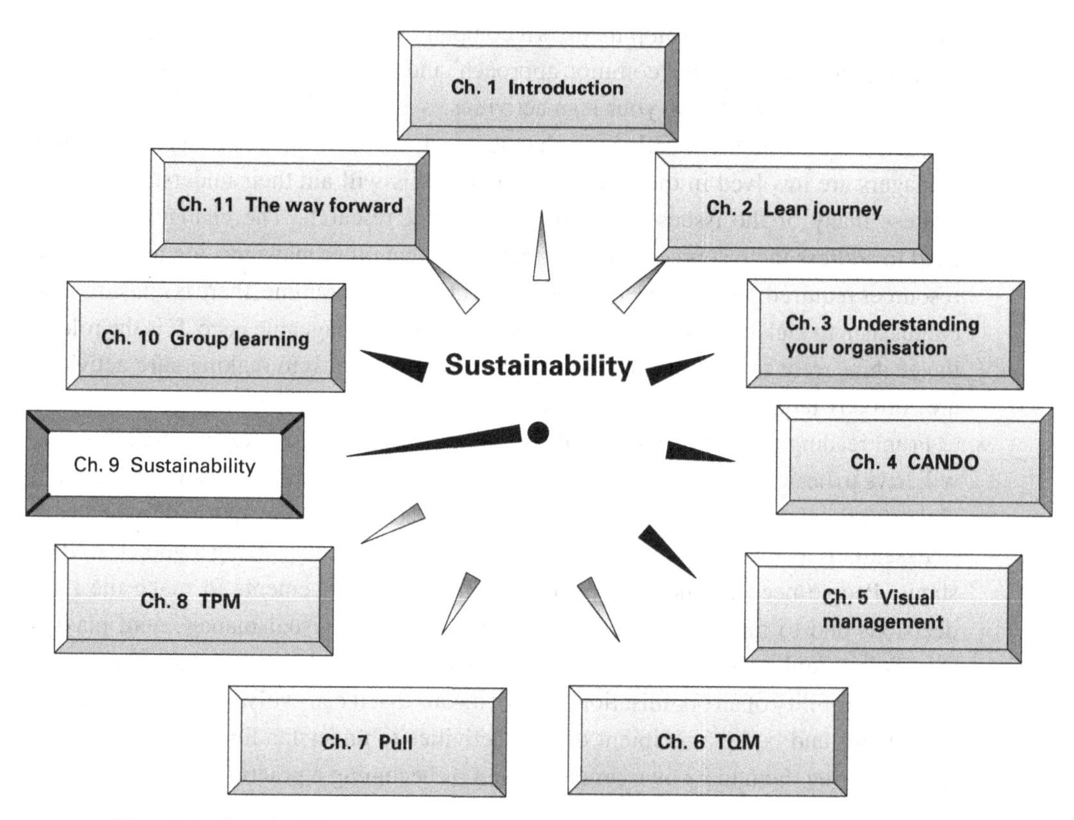

Figure 9.12 Sustainability links to other chapters

measures where appropriate, ensuring that the right activities are run to improve the VS and addressing infrastructural changes. The infrastructural changes are often missed in a lean plan and it is these that will enable the company to evolve from one state to (hopefully) a better state. Infrastructural changes tend to be in support functions, such as maintenance or finance, and they should evolve to a stage where they support the operational functions rather than impeding them. This process is often a painful one because it means considerable changes to important functions, and these changes may not always be welcome. Because of the problems in this area and the wide-ranging implications, it will often be beyond the remit of the co-ordinator, and so will have to be delegated to someone further up the management chain.

One of the infrastructural changes that the CA should initiate is the setting up of a lean office. This would hold the resources for implementing activities as well as, standardised kits such as CANDO kits and teaching and knowledge resources. The teaching and knowledge resources would be:

- "teach points" i.e. presentations focusing on one area such as visual management
- "how to" guidance i.e. how to run a CANDO activity or do a mapping exercise
- more general resources such as reference books, industrial contacts and case studies

The lean office should develop their own company's way of doing lean activities. The Industry Forum call this the 'common approach' and essentially it is applying a standard approach to the way you do your lean activities.

The co-ordinator, often called the change agent, will also have to ensure that senior managers are involved in the change process as this will aid their understanding and address many of the issues raised in the Learn 2 research. The change agent will need to address the resource issue and make sure that other managers are aware of the resources required. Often at the beginning of a change programme, there is considerable enthusiasm to kick off improvement activities at an unmanageable pace. It is the role of the change agent to control this enthusiasm and channel it into making sure activities are correctly targeted and fully realised.

From reading this chapter you will have surmised that to achieve sustainability you will have to bear it in mind when you are implementing any aspect of lean. The diagram shown in figure 9.12 shows the specific links to other chapters mentioned in the text.

Essentially you need to get your basics right and this means having a good CANDO status. People need to know why they are making improvements to make the right decisions and to maintain motivation and this is where visual management plays a role. Policy deployment sets the agenda to make sure the right improvements are being made. The ability of an organisation to learn means that it can evolve from using outside consultants and being a recipient of lean activities to understanding the principles of lean, applying them to its own processes and so becoming a practitioner of lean.

Signposts and recommendations

A PDF version of the sustainability research aimed at cell leaders can be downloaded at http://www.cf.ac.uk/carbs/lom/lerc/centre/publications/download.htm.

10 Group learning

Donna Samuel and Lynn Massey

Introduction

This final part of the book heralds something of a departure from what has preceded. Now that we have addressed the key elements of lean transformation we turn to slightly more reflective issues. In the last chapter we considered an organisation's ability to sustain change over time and this chapter addresses organisational learning, particularly in the context of a group. The final chapter outlines our reflections and deliberations at the end of the research programme and indeed at the end of the process of producing this book. Although slightly more cerebral in nature, we do try to present these matters in the down-to-earth and practically oriented style to which you have by now become accustomed.

Let us be absolutely clear at the outset about what we mean by learning and why it matters. The purpose of learning is not the accumulation of knowledge for its own sake, but rather to bring about behavioural change (Pedler *et al.*, 2003). The ability to learn new skills, to apply them and to embed them into everyday behaviour is what sustainable change is all about. It all begins with learning. For a lean transformation to successfully take place, learning is required at the operational level of the organisation, at the level of the value stream and, most definitely, at the strategic level (Dimancescu *et al.*, 1996). In the previous chapter you saw how learning determines the ability to achieve different degrees of sustainability. Ultimately an organisation's ability to learn determines its ability to adapt to changes in the environment (both internal and external) and therefore its ability to survive in today's turbulent and challenging world.

This chapter concentrates specifically on the notion of group or collective learning. It was during the early part of the Learn 2 programme that the notion of group learning appeared to some of the Learn 2 researchers to be a particularly important feature of the programme.

Essentially, the Learn 2 programme consisted of two main strands of activity:

1 One-to-one mentoring of the Learn 2 companies embarked upon lean implementation.
2 Quarterly gatherings of all Learn 2 participants including all research staff participating in the programme along with several representatives from each of the sponsor companies.

It is the second of these two activity strands on which this particular chapter of the book is focused. When the research programme was originally designed, these meetings were conceived as merely being a forum for participating individuals involved in the programme to network and share information on progress. Fairly early on in the programme, at around the third such gathering, it became clear that the meetings were developing into something far more pervasive – they had become powerful learning forums. Consequently, a few of the researchers participating in the programme were inspired to further explore the nature and extent of the group learning we were observing. In order to do so, two main research activities were conducted: first, a review of the relevant literature and, second, some primary research in the form of interviews with key individuals from participating companies in order to gain their first-hand views of how the quarterly gatherings were shaping their lean implementation journey. This chapter describes the main outcomes of this research activity.

You may be considering skipping over this chapter in the belief that its content will be of limited relevance to you and your lean implementation journey; if so, think again. It may well be that a network of like-minded organisations, brought together in a learning network, is a resource that is currently not available to your organisation. However, a small amount of time invested in following through the argument made in this chapter may inspire you from a position of dismissal to one of resolve to make sure that such a network *is* available. Indeed, the few minutes invested in reading the chapter may lead you to become a change initiator, resolved to set up a similar such network as an external resource for you to tap into and one that may be a critical element of a lean implementation journey! Remember, business improvement is often a lonely and thankless task and no man is an island.

Knowledge flows within the Learn 2 programme

You could be excused for thinking that the learning referred to in the phrase, 'learning forum', resides with the individual representatives (the lean agents and others) from the participating companies. However, our experience of these meetings led us to believe that there were many other learning experiences going on within these meetings as well. The simple diagram in figure 10.1 identifies the full extent of the learning flows within the network, much of which was initiated and crystallised at the quarterly gatherings.

In other words, while these regular meetings were of primary benefit to the participating companies, other Learn 2 participants were learning at these meetings as well. For the Learn 2 research team, for example, the programme offered the opportunity to directly observe a group of relatively disparate manufacturing organisations (outside of the automotive sector, the original 'home' of lean) in their efforts to realise lean implementation. This enabled the researchers involved to collectively debate what appears

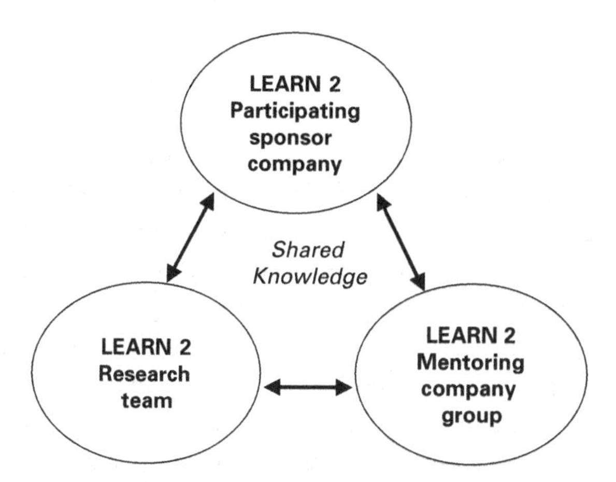

Figure 10.1 Primary knowledge flows within the Learn 2 research programme

to work well and why. Indeed much of the result of this collective debate forms the content of this book.

For the mentor companies and other external participants invited as hosts, presenters or guests to the meetings, the meetings often provided a source of new information or an alternative perspective on current knowledge.

However, the primary learning clearly happened for the representatives from the participating companies themselves. The regularity and informality of the meetings meant that they became highly powerful forums for the horizontal cross-pollination of ideas, as well as the vertical transfer of knowledge from LERC research staff and other mentors to participating companies. Later we will identify what participating companies themselves had to say about the value of these meetings. First, however, we will take a brief look at the literature relevant to the notion of collective learning.

Key findings of the secondary research

In order to crystallise our understanding of the notion of collective learning, it was considered prudent to conduct a review of the literature that seemed likely to be relevant. For those of you that have had the dubious pleasure of reading academic literature reviews before and who are therefore about to skip the following section, rest assured that the summary that follows is not comprehensive. Rather the vast body of literature referred to has been filtered down to leave a few concepts that offer valuable insights into the Learn 2 programme. But first, a few comments on the broader topic of change management in general.

Change management

In a fantastic *Harvard Business Review* (HBR) article, Jeanie Duck (1993) observes that, for most executives, managing change is unlike any other managerial task they have ever confronted. She reckons the problem is the legacy of Fredrick Taylor and scientific management. In other words, we are using a mechanistic model. We break change into small pieces and then manage the pieces. But the key to managing the change effort is not attending to each small piece, but understanding how changing one element impacts the rest. Change, then, is about connecting and balancing all the pieces. She also identifies that one of the paradoxes of change is that trust is hardest to establish when you need it the most. If a company is in trouble or in the middle of a change effort, lack of trust automatically emerges as a serious barrier. Trust, in a time of change, is based on two things: predictability and capability. Both of these are underlying drivers of the lean approach. In the context of change, predictability consists of intention and ground rules and the more that leaders clarify the organisation's intentions and ground rules, the more people will be able to predict and influence what is going to happen to them. Capability means knowing how things happen as opposed to focusing on deliverables. In a lean environment, managers know that if their processes are aligned and in control, the desired results will happen. So, rather than checking on milestones and timetables, managers should ask how the work will get done.

In another landmark HBR article, Kottler (1995) identifies some of the key reasons why so many change efforts fail. The lean approach is largely based on Japanese style management and one of the things the Japanese are very much better at than us in the west is an understanding that failure is a huge opportunity for learning, thereby increasing your chances of success next time. Kottler identifies eight key reasons for failed transformational change efforts:

Error 1: Not establishing a great enough sense of urgency. This is the first step in a change initiative and it is absolutely crucial, because a change programme requires the aggressive co-operation of many individuals. However, without motivation, people will not help and the effort goes nowhere.

Error 2: Not creating a powerful enough guiding coalition. If a critical mass is not achieved early in the effort, usually nothing much worthwhile happens.

Error 3: Lacking a vision. A key task of the guiding coalition is to develop a picture of the future that is relatively easy to communicate and appeals to customers, stockholders and employees. As a rule of thumb, if you cannot communicate the vision to someone in five minutes or less and get a reaction that signifies both understanding and interest, you have not done with this phase of the transformation process.

Error 4: Under communicating the vision by a factor of ten. Without communication, and a lot of it, the hearts and minds of the troops are never captured. Incidentally, communication comes in both words and deeds. Nothing undermines changes more than behaviour that is inconsistent with words.

Error 5: Not removing obstacles to the new vision. This is another key responsibility of the guiding coalition. In the first half of a transformation, no organisation has the momentum, power or time to get rid of all obstacles, but the big ones must be confronted and removed.

Error 6: Not systematically planning for and creating short-term wins. Real transformation takes time and a renewal effort risks losing momentum if there are no short-term goals to meet and celebrate.

Error 7: Declaring a victory too soon. New approaches are fragile and subject to regression and, although celebrating a win is fine, declaring the war won too soon can be catastrophic.

Error 8: Not anchoring the changes in the corporation's culture. Until new behaviours are rooted in social norms and shared values, they are subject to degradation as soon as the pressure for change is removed. Two factors are particularly important in institutionalising change in corporate culture. First, a conscious effort to show people how the new approaches, behaviours and attitudes have helped improve performance. Second, taking sufficient time to make sure that the next generation of top management really does personify the new approach.

Finally in their HBR article that considers why change programmes often do not lead to change, Beer and Eisenstat (1990) talk about the fallacy of pragmatic change. They argue that the theory of change is fundamentally flawed. The commonly held belief is that the place to begin is with the knowledge and attitudes of individuals. Changes in attitudes lead to changes in individual behaviour, and changes in individual behaviour, when repeated by many people, will result in organisational change. Change, then, is like a conversion experience and once people 'get religion', changes in their behaviour will surely follow. The authors argue that this theory views the change process exactly backwards. Individual behaviour is powerfully shaped by the organisational roles that people play and the most effective way to change behaviour, therefore, is to put people into a new organisational context, which imposes new roles, responsibilities and relationships on them. This creates a situation that almost forces new attitudes and behaviours on people. The authors identify six steps to effective change as being:

Step 1 Mobilise commitment to change through joint diagnosis of business problems.

Step 2 Develop a shared vision of how to organise and manage for competitiveness.

Step 3 Foster consensus for the new vision, competence to enact it and cohesion to move it along.

Step 4 Spread revitalisation to all departments without pushing it from the top.

Step 5 Institutionalise revitalisation through formal policies, systems and structures.
Step 6 Monitor and adjust strategies in response to problems in the revitalisation process.

In summary, then, change management is a difficult and elusive matter. Whilst checklists such as these are useful to help us reflect, the lean approach to change is really about creating a cycle of commitment, coordination and competence. But, depending on the starting point, getting there can be a bumpy and often uncomfortable rollercoaster ride. It is important to remember and to document why you embarked on the journey in the first place and how much has been achieved since.

We turn now to the more specific issue of how organisations learn, and, in particular, what has been said about the notion of learning as a group. We identified the main bodies of literature that cover the issue of group learning as being: organisational learning, inter-organisational networking and technology transfer and we uncovered useful concepts from all three.

Organisational learning

Recently, both practitioners and academics have identified organisational learning as perhaps the key factor in achieving sustained competitive advantage. In fact, the literature on organisational learning has been expanding in an unprecedented fashion. Organisational learning is concerned with developing new organisational knowledge with the purpose of enhanced organisational performance and, therefore, insights from this body of knowledge are highly relevant to any company embarking upon lean implementation.

Theories of organisational learning owe much to the work of Argyris, Schon and Senge. Argyris and Schon (1996) distinguish between single- and double-loop learning. Single-loop learning involves the detection and correction of errors within a set of governing variables, while double-loop learning involves changing the governing variables themselves. These concepts can best be understood by thinking about a game of chess. It is true that the best chess players make their best decisions by comparing events in the game they are playing with patterns from previous games. However, the effectiveness of this strategy depends on a fixed set of rules. In the game of business, the pattern recognition that managers have built up through experience ceases to be useful when the rules of the game change.

Double-loop learning is similar to Senge's (1990) concept of adaptive learning which centres on evolutionary changes in response to developments in the business environment and which are necessary for the survival of an organisation. Successful lean implementation requires considerable double-loop learning since management must learn how to revise or even discard dominant logics. In other words, lean implementation involves the problem of unlearning. The notion of unlearning was proposed by

Hedberg (1981) who noted that it is more difficult for organisations to discard knowledge than to acquire new information. The notion of unlearning is paradoxical in that the greater the past success, the more difficult the unlearning process. It is this need for unlearning and the constant questioning of traditional managerial assumptions that makes a network forum such a useful mechanism for companies embarking upon lean implementation.

Inter-organisational networking

In recent decades there has also been unprecedented growth in corporate partnering and reliance on various forms of external collaboration. Today companies in a wide range of industries are executing nearly every step in the production process, from discovery to distribution, through some form of external collaboration. Running throughout the literature on partnering is an argument that collaboration enhances organisational learning. Our experience of the Learn 2 programme supports this view.

Lack of trust between partners and differential ability to learn new skills are just some of the obvious barriers to effective collaboration. Sources of innovation are commonly found in the interstices between firms, universities, research laboratories, suppliers and customers. The Learn 2 programme is a successful example of this type of collaboration.

In 1996 a major government-funded project was launched in recognition of the fact that the future of business is about cooperation, collaboration and learning from each other. The project identified three primary types of inter-organisational networks: supply networks, innovation networks and learning networks. Learning networks, which match most closely with the Learn 2 programme, were defined as:

> Networks where all agents participated with the primary objective of learning, in order to increase their capability in a specific technology or subject which is neither linked to a single process or product, nor associated with a particular commercial transaction.

The central belief that underpins learning networks is that organisations need to show willingness to experiment and readiness to rethink means and ends in order to realise their potential for learning. At the outset of Learn 2, it was impossible to know whether all participating companies would exhibit such a propensity. As it emerged, most did and the benefits were widespread.

There are many advantages associated with inter-organisational learning: different perspectives can introduce novel or existing concepts new to the learner, shared experimentation can reduce risks and maximise opportunities for trying out new things and shared experiences can be supportive and confirmational. However, most importantly, shared or collective learning helps explicate the system's underlying principles and patterns. It therefore provides an environment for questionning assumptions and exploring mental models outside of the normal defensive space (Kottler, 1995).

Table 10.1 *MOPS and Learn 2*

MOPS criteria for successful network	Context of Learn 2 programme
The need for a common purpose and commitment.	The common purpose for the Learn 2 participating companies is the implementation of lean management ideas. The commitment is demonstrated by payment of a network subscription fee.
Diversity of activity	The Learn 2 network includes a number of broad and diverse activities including: the development of a website, individual company mentoring, collective meetings and workshops, individual company training and collective training
Continuous monitoring and review	Within the Learn 2 network, this takes place both formally and informally. Indeed, the research described in this chapter forms part of that activity.

Inter-organisational learning, then, can provide a powerful mechanism to facilitate double-loop or adaptive learning.

Technology transfer

Technology transfer is the process through which technology moves from outside sources to the organisation. Bessant and Rush (1995) identify the importance that intermediary bodies (such as university departments, consultancies, national and regional technology centres) can have in helping organisations with this process. They argue that there are four ways in which such intermediaries perform this role:

1 The direct transfer of specialised, expert knowledge.

2 The role of experience sharing, implicitly or explicitly (rather like bees cross-pollinating between firms, carrying experience from one location or context to another).

3 The role of 'marriage broker', providing users with a single point of contact through which to access a wide range of specialist services.

4 The diagnostic role by helping users to articulate and define their particular needs.

During the 1990s the UK Department of Trade and Industry launched a new R&D initiative (MOPS: Manufacturing, Organisation, People and Systems) aimed at identifying good practice in a range of key organisational concepts. The MOPS programme identified several important criteria for a successful network. Table 10.1 lists these criteria together with a comment on their application within the Learn 2 programme.

In essence, effective networks represent complex systems in which there is considerable interaction but where the overall transformation is one of learning. The learning outcome of the network is the dependent variable. It is this that feeds back into the operation of the network and eventually decides whether or not to continue with the network.

Table 10.2 *Champions and agents understanding of lean*

Lean change champions	Lean change agents
A basket of techniques packaged around a semi-prescriptive methodology (Air Repair)	What we are trying to do is to take waste out of the business wherever it exists (Air Repair)
A structured and organised way of looking at what you do, questioning it with healthy scepticism and coming up with ways and methodologies of doing the things you are supposed to be doing, as opposed to doing the things that most of us spend our time doing – things that we shouldn't be doing! (Mornington Cereals)	The elimination of waste to become more effective in meeting the customer's needs (Mornington Cereals)
Lean is continually improving, doing things better . . . having a methodology or process in place that enables us to standardise how we do things and problem-solve to enable us to continually add value (Health Products)	Improvement, primarily in manufacturing, but it applies to any process . . . it's about making things simple and flow basically (Health Products)

The findings of the primary research

Armed with some interesting ideas from the literature review, the next step was to carry out some primary research and find out what the Learn 2 companies themselves had to say about the programme. We interviewed the lean agents from each of the participating companies and, where possible, the lean champions as well. The information gathered during these interviews is discussed in the following section. The discussion has been sorted into three key areas: how participating companies understand lean and lean implementation, their key achievements during the first year of lean implementation and, finally, their opinion of the Learn 2 programme and suggestions for how the programme might be improved.

Key area 1: Understanding lean

We wanted to know how our interviewees would describe the lean concept to us and, in particular, whether their understanding of lean had changed during the first year of the programme. Generally, lean change champions, usually senior personnel, referred to lean as a 'basket of techniques' or a 'set of methodologies', while lean change agents tended to give rather more succinct answers, often quoting more directly from the texts.

For some companies, lean is seen as part of a broader initiative with a more generic title such as 'world-class manufacturing', while for others the word 'lean' had taken on a wholly different meaning to the one found in the literature. Indeed, one company reported lean as being a 'bandwagon'. Lean for this company was seen as a vehicle through which people could become trained up and build themselves a pleasant career, but where making real or physical changes happen proved extraordinarily difficult.

> It's almost as if doing things, moving the product is dirty, hard work, people just don't go there. People's promotional aspirations, development plan . . . is to get away from the product, that is the way they get on in the company. That's wrong for a manufacturing site. (Lean agent, Medical Devices)

Regarding whether their understanding of lean had changed or developed during the course of the first year of the programme, most of our interviewees denied that it had. However, they commented that their understanding of the difficulties of implementation had increased. Several respondents admitted that they had thought it to be far more technical in its nature and had discovered that it is actually about people and behaviour:

> In the beginning I had a belief or an understanding that it was more technical and it's changed to – it's actually about people issues. (Lean agent, Air Repair)

> What I wasn't aware of is that lean has quite a number of people aspects. (Lean agent, Heavy Products)

> Our understanding of which obstacles we actually have has improved. (Lean champion, Air Repair)

> I guess my understanding of the ramifications of what's possible, the hurdles of going through it and the human element. (Lean champion, Mornington Cereals)

The findings suggest that the literature on lean manufacturing is technically focused and can therefore misguide practitioners as to the extent of ethnocentric (or people-centred) change that will need to take place for successful implementation. This confirms the Learn 2 research team's belief that it is the 'messy' and 'gritty' details of lean implementation that offer the greatest research challenge.

Key area 2: Key achievements, during the first year

We asked companies what they felt had been the key areas of achievement during the first year of their lean implementation. Their responses were diverse but often very impressive.

Air Repair had concentrated their efforts on improving the mindset of shop-floor workers, encouraging them to participate in improvement activities. Air Repair made enormous improvements in its relationship with a powerful trade union. This was seen as a necessary prerequisite to successful lean implementation as this cultural change would need to be firmly embedded before significant bottom-line improvements could be realised.

> It isn't the norm for our junior and middle management to spend their time talking to people . . . I think the sophistication of our management to understand the subtlety of what we are trying to do is probably one of our biggest weaknesses . . . If it's three days since someone talked to them about lean, then management's given up on lean. (Lean champion, Air Repair)

Heavy Products concentrated on the production flow, reconfiguring the supply system so that strategic stock was held at a different point within the internal supply chain. The company deliberately focused less on the workforce mindset change and more on the redesign of their production system. They reported that they had halved the inventory within the particular value stream they had addressed. As for difficulties, this company maintained that there had been few, primarily because of the modest approach that had been adopted.

We're not trying to wallpaper the whole room at once. So we're not being seen to be doing things unnecessarily, just for the sake of doing things . . . we've been fairly selective about what we have done. (Lean agent, Heavy Products)

Health Products focused its efforts on a highly participative CANDO, which they reported had been received by the workforce with unprecedented zeal and enthusiasm.

The response we got from the people in production was fantastic. The night shift in particular had never got involved in anything before. (Lean agent, Health Products)

The CANDO programme was very successful because the employees embraced that. (Lean champion, Health Products)

Our interviewees could only surmise at why this was the case. It could be the fact that so many individuals were involved in the programme (perhaps unlike previous improvement initiatives); alternatively, it could be the extent of communication and explanation that took place before and during the programme (again, something that may have been underestimated in previous improvement efforts). However, they maintain that CANDO is a technique that directly improves the working environment and, as such, is highly relevant and beneficial to the workforce. As for benefits to the company, each production cell targeted for CANDO showed a significant improvement in both yield and output, in some cases the improvement was as much as 25%. Health Products has now assigned two dedicated full-time resources to sustaining the improvements made through their CANDO programme and expanding the technique to other areas of the business. The same company is currently planning to enter this year's 'Best Factory' award.[1] As for difficulties, this company suggests that trying to follow something as engaging as CANDO is not easy.

Mornington Cereals focused their efforts on process improvement and have seen significant improvement in their OEE[2] performance and in their profit and loss account. Mornington identified resourcing improvement activities as their main difficulty, along with the wider problem of cultural change.

It's difficult to get across to other managers who have other priorities, other big problems, that this is really important. (Lean champion, Mornington Cereals)

[1] The Best Factory Award is a Department of Trade and Industry initiative that recognises and celebrates excellence in UK manufacturing.

[2] OEE stands for overall equipment effectiveness and is a high level measure often used in lean plants. It relates directly to the six big losses and is calculated by availability × performance rate × quality rate.

As for how lean agents viewed their personal role within the lean implementation process, words such as 'disciple', 'oracle', 'conscience' and 'facilitator' emerged. Most lean agents rose to the challenge with enthusiasm, though one commented on the steepness of their own, personal learning curve and another pointed out the problem of implementing improvement activities without line responsibility and ownership.

Key area 3: Opinions of the LEARN 2 programme and suggestions for improvement

All interviewees heralded their participation in the programme as a success though in varying degrees and for varying reasons:

> The short period of time involvement has been far more successful than just about any other project we have tried to do. (Lean agent, Air Repair)

Some companies were not as far along with their lean implementation as they had originally hoped to be, progress had been slower than they had anticipated. However, earlier comments on their increased understanding of the difficulties of implementation suggest that they may initially have been rather over-optimistic. One interviewee described the programme as a 'win–win' scenario,

> You get to experiment on us . . . the win for us is that you get to hold our hand and provide some resources that allow us to build up skills within our business to allow us to get going. To me there is a win–win. (Lean agent, Mornington Cereals)

With the exception of one company, all had been highly active in the networking element of the programme. The benefits derived and the way in which the network was used varied considerably. However, the main benefits are summarised below, embellished with quotations from the interviews.

1 The network exposes companies who are struggling with change implementation to others who are in a similar situation

One respondent commented that:

> Working with some non-competitive business and getting some good alignment and openness between practitioners in each of the businesses is extremely powerful. (Lean agent, Mornington Cereals)

Another described the network as both a 'comfort blanket' and a learning opportunity,

> At the outset we recognised it as being a marathon as opposed to a sprint . . . the thing about marathon improvement programmes is that you do need a 'shot in the arm' every so often to remind you that it is still worth going. And I think having a forum and a network where we can share other people's successes and failures and disappointments and everything else, we felt was going to be quite important

for the medium term sustainability . . . I think the different industries are particularly helpful to us. (Lean agent, Air Repair)

Being part of a group of people who are going down a learning curve together. (Lean champion, Air Repair)

The exception, in other words the company that had not actively participated in the network activities, held the opposite view. Their reasons were time constraints, lack of clarity as to the benefits and resultant lack of motivation. While most of our companies regarded the diversity and non-competitive nature of the network as its strength, this company would have been more motivated had another similar company been involved.

We've had very limited involvement in terms of the network . . . I'm not absolutely clear in my own mind what the potential benefits are, perhaps because the industries involved are extremely different from us . . . (Lean champion, Heavy Products)

One of the companies recognised that the network was an opportunity to 'tap into other companies', but felt that they were the leaders and that their own rate of progress was being stifled by the group.

The network helped. It was good for our guys to go and present to different companies . . . what we recognise is that we are the leaders in some regards rather than the followers . . . we want to work with comparative companies who we can benchmark ourselves against, not discover that we are leading the way and other groups are learning from us and our progress is suffering as a result of it. (Lean agent, Health Products)

2 The network provides a 'catalyst for action'

The Learn 2 companies are required to formally report on their progress since the last network meeting. One respondent compared this with the impact of external auditors, suggesting that the network provides a 'catalyst for action'.

Companies commented that the requirement on them to deliver a formal presentation on progress urged them to ensure that there was progress to report!

3 The network provides an opportunity for internal teambuilding and motivation

One company used the network meetings as a forum for internal teambuilding.

One of the benefits would be that we have always tried to have the core team attending. We have used it as a teambuilding thing. So that the meetings were used as a way for us to meet outside of work. (Lean champion, Health Products)

Another interviewee commented that he used the network meetings in two key ways: first, as a way of keeping his internal lean implementation team's 'batteries topped up';

Table 10.3 *Benefits*

Benefit derived	Scores: 5 = strongly agree 4 = agree 3 = indifferent 2 = disagree 1 = strongly disagree
1 Meeting and general networking with other companies going through a similar change programme	One respondent scored 3, another 4 and the remainder scored 5.
2 Development of understanding of lean (through presentations by research staff at LERC)	One respondent scored 5, the remainder scored 4.
3 Development of understanding of lean implementation (through presentations by other network members)	Three respondents scored 3, the remainder scored 5.
4 Learning from external network participants (e.g. Learn 2 network host companies or Learn 2 mentor companies)	Three respondents scored 4 and the remainder scored 5.

second, to help broaden the views of some in the organisation who might be cynical towards lean, in other words, potential 'change inhibitors'.

We've taken people from the shopfloor along. At the last one we took a Union representative. It ends up being a good way to touch some people and get them motivated. (Lean champion, Air Repair)

Companies also reported several more tangible, offshoot business opportunities that had resulted from their participating within the network. Health Products identified an opportunity to sell their products through one of the major retailers (who participates in the network as a 'mentor' to the group). This opportunity was passed over to their marketing department and will hopefully lead to a concrete outcome in the future. The same retailer has one of our network companies, Mornington Cereals, as its supplier. Individuals from these two companies met as a result of their mutual participation in the Learn 2 network, formed a relationship, and have identified various anomalies and waste removal opportunities within their supply chain.

In order to capture relative data, interviewees were asked to score the extent of their agreement on a scale of 1 to 5 against a list of benefits believed to be present. In terms of how the network could be improved, our interviewees offered several suggestions.

One respondent suggested that a separate meeting of change champions, often the MD, would be useful. This meeting would be less concerned with the gritty details of lean implementation and more a forum for the cross-fertilisation of ideas and experiences. Some companies felt that LERC should be harder 'task masters', offering public comments and feedback on progress to date.

I think that formal feedback may be painful but nevertheless useful. (Lean champion, Air Repair)

In conclusion

This chapter describes some of the findings of a series of interviews conducted with each of the Learn 2 participating companies at the end of the first year of the programme. The purpose of the interviews was to understand the views of each of the companies regarding lean implementation generally and, more specifically, their participation in the Learn 2 programme. The findings suggest that participating companies are generally positive towards their involvement in the Learn 2 programme. Indeed, most of our interviewees offered testimonials to the many benefits of group learning. Whilst these benefits can be difficult to quantify or even articulate, they can be broadly summarised in the following way:

1 The programme allowed participants to tap into tacit expertise that resides within a University.
2 The programme allowed participants to draw both knowledge and comfort from others who are going through a similar difficult implementation process.
3 The programme provided a mechanism for internal team-building and training.
4 The programme provided a catalyst for action enabling companies to overcome the inertia that often follows the initial flurry of activity and enthusiasm that often comes with major change programmes.

Lean implementation requires companies to question and revisit many of their basic assumptions. In the language of learning theories, it requires 'double-loop' or 'generative' learning. More latterly theorists have referred to a further type or level of learning, which involves a fundamental organisation paradigm shift as organisations begin to 'learn about learning' (Pemberton and Stonehouse, 1999). It is our belief that the Learn 2 programme provided a forum that allowed participants to begin to experiment with this notion. The fact that participants have to report on their lean implementation progress on a regular basis to an external organisation forces them to revisit and reflect on their activities. In this way they constantly have to monitor their own performance and behaviour and to address their failures. Addressing failure is one the keys to unlocking implementation success. One author makes an insightful distinction:

A productive failure is one that leads to insight, understanding, and thus an addition to the commonly held wisdom of the organisation. An unproductive success occurs when something goes well, but nobody knows how or why. (Garvin, 1993)

11 Reflections and future challenges

Nicola Bateman, Ann Esain, Lynn Massey, Nick Rich and Donna Samuel

We have addressed the subject of lean holistically, though our discussion has by no means been exhaustive. Indeed there are endless aspects of lean implementation we could have tackled: the application of lean into the new product process, supply chain management or translation of lean into a service environment, to name just a few. However, for the purposes of focus and keeping our target readership, we have limited our discussions to lean implementation in a manufacturing environment. In this chapter we do allow ourselves a little latitude, and cast our thoughts to what we believe are some of the major challenges facing our target readership today. An awareness of these issues is a sound starting point for solving them. Key to this is the prime driving force of lean – that of delivering value to customers. Before we continue, however, we offer the reader a brief summary of this journey through the book so far. For those that have read the book from end to end, this will act as a review and reminder; for those that have dipped in and out, this will offer a useful overview.

Brief review of the book

In our Introduction we present our research sponsors and dismiss the myth that lean is just an automotive or Japanese phenomenon. We identify our target readership and set our aims and objectives in writing this book. Chapter 2 sensitises the reader to the origins of lean thinking and introduces the basic principles, pillars, and assumptions that underlie the lean approach. Chapter 3 compares and contrasts the traditional and lean business models and allows readers to position their own practices against these stereotypes. The transition from traditional to lean is an evolution but may be guided by knowing what the 'end state' looks like. This chapter provides an outline of the key practices and aspects of effective change processes and then goes on to explore the process of policy deployment – the guiding hand of lean implementation.

The next part of the book examines more specific aspects of lean. Chapter 4 outlines the common starting point and essential building blocks on which other lean techniques can be built. The CANDO process is a major sign of visible change in the factory and a means for everyone to make a demonstrable difference. It provides a way of mobilising the creative talents of the workforce, a great untapped resource of any organisation, with immediately apparent results. The visual management chapter (5) is closely related to CANDO but is a subject of lean change that is poorly explored by most texts. Visual management is designed to make life easier and to simplify channels of communication. It creates a shared vision and enables teams to assess their performance and understand what is expected of them. Chapter 6 explores quality systems and the modern evolution that is Six Sigma. It is a return to the design and management of the core operational system – quality management. At this stage, the business in transition will look to eliminate further waste. The pull chapter (7) provides an account of the material flow and delivery process options available to allow high-quality materials to flow in an even more controlled manner. Once again, myths have been dispelled that kanban is a mandatory component of lean production. At this point in the lean evolution, the company production system becomes formalised and hard-wired. Finally in this part, Chapter 8 addresses Total Productive Maintenance (TPM) – a badly understood system of management, which is grossly portrayed by many texts as just operator involvement and the measure Overall Equipment Effectiveness (OEE). TPM is the means of making processes robust and ensuring that machine performance supports flow production and bufferless operations.

In the final part, we take a slightly different direction and address some broader issues. Chapter 9 outlines the critical concept of sustainability of change – an elusive aspiration of contemporary management. It examines sustainability issues at multiple levels of the organisation to highlight the importance of aligning improvement events and the timing of changes in the value stream with the challenges of the company-wide policy deployment approach. The group learning chapter (10) examines the benefits accrued from the Learn 2 programme and explores the importance of collective learning and experience sharing beyond the organisation. This chapter reviews the comments, reflections and personal points of learning of the managers involved in the network and how they 'made sense' of their, and others, learning processes.

Throughout our book we intended to take a systems approach to lean change rather than presenting a collection of tools. We aimed to synthesise the repeated pattern of behaviours that we have observed over time and have distilled the key learning points that we hope will assist you in developing a logical approach to your own lean transition. It is all too easy to abdicate the responsibility for designing the future organisation and its armoury of improvement techniques in favour of buying in consultancy help. But often, paying for such assistance, managers fail to reason through the true issues affecting the design of the firm. It is imperative that lean change champions appreciate that lean business design and its principles were not handed down from the mountain on tablets of stone – their application needs to be discussed and worked out by the entire management collective and can only be implemented by a leadership with commitment and a constancy of purpose.

It will have taken you far less time to read this book than to reach the point of sustaining lean within your business. Both journeys should have been enjoyable and served to focus your mind, and those of your colleagues, on adding value by eliminating waste. Your lean journey will have, in the beginning, made you painfully aware of the problems in your factory, taken you through many stages of control to bring stability to your workflow and finally added the necessary elements of sustainable process improvement. Taking part in this journey will have started the development of a learning environment. The operators are involved in the day-to-day continuous improvement activities, and are a valuable resource for informing and contributing to the decision-making process. The development of skills in problem solving relieves middle managers, allowing them to focus on more strategic aspects of the lean journey. That is not to say the process of lean implementation is easy or without barriers, but, given a logical process of implementation and an attention to the human side of learning, how to improve the factory can become a stimulating and rewarding environment. Contrast this with the traditional stereotype of a rather bleak, dirty and chaotic workplace that is full of danger and stress. Without the process that has been described in this book, the traditional stereotype is likely to live on and will do little to promote manufacturing as a 'competitive weapon'. In reality though, most factory operations areas are vibrant workplaces, despite popular belief and media images.

Sustaining the lean drive

Even after 60 years of experimentation and dissemination of lean practices to the supply chain, Toyota and its suppliers remain humble and will often state that there is much still be done and achieved. In reality there always will be. Products will change, new ways of doing business will be developed and new technologies will emerge. Cars and the many other products/services which make up Toyota and the competitive environment for us all will always be dynamic and pose new threats and opportunities.

The achievements of the Learn 2 Project of research work have continued, and each participating company has equipped itself with the learning processes and lean techniques needed to sustain these improvements. Despite these achievements, there is no room for complacency – examining your success is good, but arrogance can blind you to further improvements. So you can never relax your lean initiative and, when times get hard, you must double your efforts and not fall into the old trap of cutting training budgets and wondering why improvements come to a grinding halt. Even if you accept there is room for improvement at your factory, you must also keep a keen eye on the many clouds upon the horizon which pose new challenges to be tackled. We will now review some of these issues to help you craft your future agenda.

The challenges to lean producers and lean professionals

The first major challenge that faces most lean companies is to grow and grow. It is no wonder that Toyota retains a corporate policy of lifetime employment for their number one asset – the employees. To achieve this, with an ever-increasing average age of worker, is no mean feat, but is one that can be only be achieved by continually reinventing the corporation. Growth is key to lean success (whether by stretching products or by diversification) and, without it, improvement activities will release labour which at some point in the future cannot be employed elsewhere to chase waste. Growth means new products and new challenges. It means the glass ceilings of promotions are removed and brings with it the discomfort of the new but also the excitement of building something again. This typifies Toyota, from looms to cars to forklifts to all manner of ventures, including construction, where the pursuit of profitable growth has been common. This is the challenge that scares us most in the west with our saturated markets and under representation in most of the important emerging industries of the future.

The time-to-market challenge

One of the common denominators in industry today is the measurement of time as a key form of competitive weapon. That is, the time to fulfil orders now and also the time to get ideas and concepts through a robust design process and to the market. For most lean authors, this time is characterised by the process of order fulfilment; however, the most important aspect for the future is the design of new products and commissioning of new processes, to become 'first in market' or a very close second, which adds to the viability, sustainability and growth. Second in market is not necessarily a bad position, often it is better to use the design process capability as a means of letting another business create a market and for the business to sell into an expanding market without the development costs. Whichever approach is favoured, the issue is getting 'valued products' to new and existing customers in shorter and shorter times.

During your lean journey you may have been involved in supplier development activities. The knowledge that has been generated from this allows all parties to understand the value chain as a whole system, and, as a consequence, areas where additional value can be created are more easily determined. This helps when redesigning the system to generate the most benefit for the supply chain as a whole. These skills are a key organisational asset, contributing to the early stages of product development or product redesign and prototyping. Your lean journey, using cross-functional activities, will have helped break down functional silos and will have helped to develop a consensus process approach to product development. It will also have helped you to share and retain the knowledge within the value chain, ensuring that the lessons learned from past experiences are transferred to those who will benefit, thereby helping to reduce reworking during product development.

The second aspect of the design process concerns that of re-design and taking existing products through processes of value engineering to find ways of enhancing value, whilst reducing costs or engaging in late customisation/configuration. It is true that, for all of the lean preaching over the last 20 years in the West, it is still possible – even probable – that many businesses are making a 'bad product' in the best lean way. As such there remains the challenge to develop greater links with customers and to 'lock them into' the designs of future products in a way that makes them an owner of, and dependent upon, the design process. Design is a key capability and one which means managers must understand processes such as Research and Development if businesses are to take advantage of the new materials and technology breakthroughs which reside elsewhere, often inside the labs of Universities.

The supply challenge

The concept of being just one part of a bigger supply chain was first talked about in the mid 1980s, at about the same time that Just-In-Time (JIT) logistics and the study of the integrated supply networks of Japanese manufacturers were first being noticed by Western management authors. For cars, over 80% of the vehicle is made and transferred from suppliers and the same is true for many other sectors. This creates a new opportunity for lean businesses that have spent time levelling the orders they receive from customers and level loading the factories production to create a standardised and repetitive manufacturing schedule. For many businesses, interacting with suppliers still remains an area of untapped potential and therefore an opportunity for the future. In an era where skilled supply management people are in relatively short supply, the importance of this aspect of lean management cannot be underestimated. Suppliers are a critical source of innovation. As such, the lean lessons 'inside the works' need to be extended to the supply base. No matter how severe the penalty charges for late or non-delivery, you are dependent upon your supply base and you are responsible for the correct design of your supply base. No supplier can ever know enough about a customer in the same way that you can never know enough about your customers and consumers. Holding suppliers at 'arms length', as we have done in the past, and demanding the introduction of lean practices at suppliers, even though they were not practiced in-house, is not a viable strategy. Toyota mastered her internal production system and smoothing of demand long before taking it outside the supply base to introduce Just-In-Time logistics systems. This latter activity, that commenced in the 1970s and continues to this day, involves regular meetings of supplier groups, strategy and 'best practice' sharing. It is a challenge which is now presented to manufacturers in the UK and an issue which will challenge the business in terms of local supply or long-distance supply from cheap labour sources. It is a debate well worth having – there is considerable competitive advantage to be gained from effective supply chain management.

The skills vacuum

For the UK and most of Western Europe, our education systems are simply not producing as many engineers or engineering-skilled workers. This reflects a growing trend away from selecting engineering as a career, since there are a myriad of alternative careers and sectors all competing for the same pool of talent and many of the others offering more attractive remuneration packages. Whatever you believe and however you assign fault is irrelevant – there are not enough skilled people around. Here the challenge is therefore to recruit for this scarce resource and to train up other employees. This challenge is a hard one – traditional and 'time-served' employees who are coming to retirement age create a vacuum and you must understand how you are going to replace these skills. The most likely course of action is to up-skill and that means the re-education of the workforce, maybe via National Vocational Qualifications (NVQs). Taking this option is not easy – it means time away studying whilst the current talent is stretched further, but you must look to the long term. Without these skills, the future is gloomy. One thousand kaizen events will result in nothing if your engineering skills are low and correct countermeasures are not introduced and that calls for good engineering awareness.

Environmentalism

As corporate responsibility is now a key theme for managers and different stakeholders put pressures on businesses to support the local community and deal fairly with supply sources, businesses that buy from 'dubious' sources are being exposed by lobby groups for their activities, and brand names have been tarnished by campaigns to 'out' businesses seen as 'doing wrong'. Just take the examples of sports footwear and the criticisms that suppliers use child labour or work in conditions that are dangerous and inhumane. These issues are important for any business. If customer value switches to supporting the environmental or the 'social justice' lobby, then this will have a big impact on the willingness of consumers and customers to select businesses which buy indiscriminately.

Globalisation

Never far from the strategic agenda and an aspect of the issues that have already been raised is the concept of increasing globalisation and more competition against 'low cost' producers. To hold on to existing accounts you must make your customers dependent on you so that it just does not warrant the time and effort to switch sources to your competitors. So your attention to 'sustainability' cannot waiver – instead as times get tough these budgets should be doubled and not halved. To compete you must be better than the rest and differentiated in terms of the 'value offering' you put to your chosen markets. In fact, these pressures existed at the beginning of our Learn 2 programme and have intensified at the end. In response, our sponsors have not stopped their programmes of lean implementation and extension, but increased them. The initiative and its many years of pay backs to the capability and performance of the firm have therefore been reinforced. Globalisation has been the single biggest source of debate for our sponsors and it is likely to be yours too

Final words

The contents of this book have been written by a team of researchers who have spent the majority of their working lives involved in change management and improvement activities. This experience has been achieved through working with a wealth of companies located throughout the world. We hope that the book gives you some key pointers in carrying out a lean transition and that it triggers ideas that provoke discussion at all levels, but most importantly at senior management and corporate levels. We hope this book has given you some of the enjoyment we have experienced during our research and that you are now in a better position to 'craft' the future of your business. We congratulate and wish luck to any organisation and individual who takes on change. It is never easy to challenge the status quo and this bravery should be commended.

Finally, we would like to take this opportunity to thank the Learn 2 companies, their change agents and the many other individuals at all levels of the organisations we met for sharing their learning experiences with us. Their role in helping us understand the barriers and benefits has been invaluable.

References

Alford, D. (2002), *Single Point Lesson*, Maesteg: Cosi Limited.

Argyris, C. and Schon, D. A. (1996), *Organizational Learning II*, London: Addison-Wesley.

Bateman, N. (2001), 'Sustainability: a guide to process improvement', LERC, Cardiff University, Cardiff.

Beer, M. and Eisenstat, R. (1990), 'Why Change Programs Don't Produce Change', *Harvard Business Review*, November–December.

Bessant, J. and Rush, H. (1995), 'Building bridges for innovation: the role of consultants in technology transfer', *Research Policy*, 24.

Bicheno, J. (2000), *The Lean Toolbox*, Buckingham: Picsie Publications.

Bicheno, J. and Catherwood, P. (2004), *Six Sigma and the Quality Toolbox*, Buckingham: Picsie Books.

Deming, W. E. (1986), *Out of the Crisis*, Cambridge, MA: Center for Advanced Engineering.

Dimancescu, D. (1992), *The Seamless Enterprise,* New York: Harper Collins.

Dimancescu, D. Hines, P. and Rich, N. (1996), *The Lean Enterprise: Designing and Managing Strategic Processes for Customer Winning Performance*, New York: Amacom.

Duck, J. (1993), 'Managing change: the art of balancing', *Harvard Business Review*, November–December.

Feigenbaum, A. (1991), *Total Quality Control*, New York: McGraw Hill.

Garvin, D. (1993), 'Building a learning organisation', *Harvard Business Review*, July–August.

George, M. (2002) *Lean Six Sigma: Combining Six Sigma Quality will Lean Production Speed*, New York: McGraw Hill.

George, M. and Rowlands, D. (2003), *What is Lean Six Sigma?* London: McGraw-Hill.

Goldratt, E. (1984), *The Goal: A Process of On-Going Improvement*, North River Press.

Goldratt, E. (1986), *The Race*, London: Gower Publishing.

Goldratt, E. (1991), *Haystack Syndrome*, North River Press.

Goldratt, E. (1993), *The Goal*, London: Gower Publishing.

Goldratt, E. (2002), *It's not Luck*, London: Gower Publishing.

Grief, M. (1991), *The Visual Factory*, Portland Or: Productivity Press.

Gross, J. and McInnis, K. (2003), *Kanban Made Simple: Demystifying and Applying Toyota's Legendary Manufacturing*, New York: AMACOM.

Harrison, A. (1992), *Just In Time in Perspective*, London: Prentice Hall.

Hartmann, E. (1992), *Successfully Installing TPM in a non-Japanese Plant*, New York: TPM Press.

Hedberg, B. (1981), *How Organisations Learn and Unlearn*, in Nystrom P. C. and Starbuck H. (Eds), *Handbook of Organisational Design*, London: Cambridge University Press.

Hill, T. (1985), *Manufacturing Strategy*, London: Macmillan.

Hines, P. and Rich, N. (1997), 'The seven value stream mapping tools', *International Journal of Operations and Production Management*, **17**(1): 46–64.

Hines, P. and Taylor, D. (2000), *Going Lean: A Guide to Implementation*, Cardiff Business School.

Hirano, H. (1996), *5 Pillars of the Visual Workplace: The Source Book for 5S Implementation*, Portland, OR: Productivity Press.

Imai, M. (1986), *Kaizen*, New York: Free Press.

Kolb, A. (2000), *Punished by Rewards*, New York: Houghton Mifflin.

Kottler, J. (1995), 'Leading change: why transformation efforts fail', *Harvard Business Review*, March–April.

Lamming, R. (1993), *Beyond Partnership: Strategies for Innovation and Lean Supply*, London: Prentice Hall.

Louis, R. (1997), *Integrating Kanban with MRPII*, Portland, OR: Productivity Press.

Lu, D. (1986), *Kanban and JIT at Toyota*, Portland, OR: Productivity Press.

Maggard, B. (1992), *TPM that Works*, New York: TPM Press.

Mather, H. (1988), *Competitive Manufacturing*, Englewood Cliffs: Prentice Hall.

McCarthy, D. and Rich, N. (2004), *Lean TPM*, London: Elsevier.

Monden, Y. (1983), *The Toyota Production System*, Portland, OR: Productivity Press.

Monden, Y. (1993), *Toyota Management System*, Portland, OR: Productivity Press.

Monden, Y. (1998), *Toyota's Production System*, Portland, OR: Productivity Press.

Nakajima, S. (1988), *Introduction to TPM*, Portland, OR: Productivity Press.

Ohno, T. (1988), The *Toyota Production System: Beyond Large-Scale Production*, Portland, OR: Productivity Press.

Ohno, T. (1988), *Toyota Production System: Beyond Large-Scale Production*, Portland, OR: Productivity Press.

Pedler, M., Burgoyne, J. and Boydell, T. (2003), *A Managers Guide to Leadership*, London: McGraw Hill.

Pemberton, J. and Stonehouse, G. (1999), 'Learning and knowledge management in the intelligent organisation', *Participation and Empowerment: An International Journal*, **7**(5).

Peters, T. and Waterman, R. (1982), *In Search of Excellence*, New York: Harper & Row.

Rath and Strong Consulting (2004), *Pocket Book for Advanced Six Sigma Tools*, London: McGraw Hill.

Repenning, N. and Sterman, J. (2001), 'Nobody ever gets credit for fixing problems that never happened', *California Management Review*, **43**(4): 64–88.

Rich, N. (1999), *TPM: The Lean Approach*, Liverpool: Liverpool Academic Press.

Rich, N. (2001), 'Turning Japanese?', Ph.D. Thesis, Cardiff University.

Rich, N., in Taylor, D. and Brunt, D. (eds) (2000), *Manufacturing Operations and Supply Chain Management – The Lean Approach*, London: Thompson Business Press.

Rother, M. and Shook, J. (1998), *Learning to See*, Lean Enterprise Institute.

Sekine, K. (1992), *One Piece Flow*, Portland, OR: Productivity Press.

Senge, P. (1990), *The Fifth Discipline: The Art and Practice of the Learning Organization*, London: Century Business.

Senge, P. (1993), *The Fifth Discipline*, London: Century Business.

Shingo, S. (1986), *ZQC: Source Inspection and the Poka Yoke System*, Portland, OR: Productivity Press.

Shingo, S. (1988), *Non-Stop Production: The Shingo System for Continuous Improvement*, Portland, OR: Productivity Press.

Shingo, S. (1989), *Study of the Toyota Production System: From an Industrial Engineering Viewpoint*, Portland, OR: Productivity Press.

Shirose, K. (1996a), *TPM Team Guide*, Portland, OR: Productivity Press.

Shirose, K. (1996b), *TPM for Every Operator*, Portland, OR: Productivity Press.

Shirose, K. (1997), *Focused Equipment Improvement for TPM Teams*, Portland, OR: Productivity Press.

Spear, S. and Bowen, H. (1999), 'Decoding the DNA of the Toyota Production System', *Harvard Business Review*, Sept.–Oct.

Suzaki, K. (1987), *New Manufacturing Challenge*, New York: Free Press.

Tajiri, M. and Gotoh, F. (1999), *Autonomous Maintenance in 7 Steps*, Portland, OR: Productivity Press.

Walton, M. (1992), *The Deming Management Method*, London: Mercury Publications.

Willmott, P. and McCarthy, D. (2000), *TPM: A Route to World Class Performance*, London: Butterworth Heinemann.

Wireman, T. (1992), *Inspection and Training for TPM*, New York: Industrial Press.

Womack, J. and Jones, D. (1996), *Lean Thinking*, New York: Simon & Schuster.

Womack, J., Jones, D. and Roos, D. (1990), *The Machine that Changed the World*, New York: Rawson Associates.

Index

Aisin Corporation, 141

Bateman, Nicola, 163
Bessant, John, 192
Bicheno, John, 59, 104, 107
Bottleneck, 29, 74, 135, 156

Cando, 8, 27, 54, 57, 60–62, 63, 86, 90, 93, 101, 102, 103, 155, 158, 160, 167, 170, 175, 176, 178, 182, 183, 184, 195
  5S, 28, 60, 63
  Cando, Stage 1, 65–66
  Cando, Stage 2, 63–65
  Cando, Stage 3, 66
  Cando Stage 4, 67, 68
  Cando, Stage 5, 68–69
  Cando Organisation, 69, 73, 74, 76
  Red Tags, 64, 102
Catherwood, Phil, 104
Cellular Manufacturing, 130
Change Process, 189–190
Change Agents, 52, 163, 172, 175, 176, 177, 178, 193
Change Champions, 52
Communications Board, 83, 85, 87, 88, 89
Competition, New Rules of, 11–13
Competitive Threats, 11–12
Constant Work In Process (CONWIP), 135
Cross Functional Management (CFM), 55, 56

Define, Measure, Analyse and Verify (DMAV), 110
Define, Measure, Analyse, Innovate, Control and Transfer (DMAICT), 110
Deming, Dr. Edwards, 39, 54, 81, 100, 102, 104, 110
Denso Corporation, 141
Department of Trade and Industry (DTI), 79, 192, 195
Dimancescu, Dan, 32, 59, 60, 185
Duck, Jeanie, 188

Early Equipment Management (EEM), 153, 161
Failure Modes Effect Analysis (FMEA), 100, 102
Fire-fighting Syndrome, 2
Ford Motor Company, 102, 124

Gauge R and R, 114–119
General Electric (GE), 110
George, Michael, 96, 110
Grief, Michael, 80, 81, 84

Hartmann, Edward, 147
Harvard Business Review (HBR), 59, 188, 189
Heijunka Board Levelling, 127, 128
Hidden factory, 95
Hill, Terry, 95
Hines, Peter, 48, 96
Hirano, Hiroyuki, 65, 79, 82
House of Lean, 26, 28

IDEA, 32
Imai, Masaaki, 147
Investors In People, 54

Japanese Institute of Plant Maintenance JIPM, 141, 145
Jones, Dan, 5, 11, 24, 31, 105, 110, 125, 143
Just In Time (JIT), 39, 141, 202, 204

Kamikaze Kaizen, 33
Key Performance Indicators (KPI), 41, 87
Kolb, David, 84

Lean Enterprise Research Centre (LERC), 3–4, 163, 164, 165, 187, 199
Lean, Implementation Logic, 27–30, 47–54, 57, 103
Lean, Five Principles, 15, 16, 19, 20
Lean, Journey, 18, 19
Lean Misconceptions, 19, 23
Lean Production, 1, 13, 15, 29, 32, 39, 47, 90, 147
Lean Promotion Office, 24, 28, 51, 54, 184
Lean Thinking, 13–18, 31, 45, 110, 111, 119, 134
Lean Toolkits, 2
Learn 2, 52, 82, 89, 90, 91, 121, 163, 173, 180, 185, 186, 191, 193, 194, 199, 202

Management, Role of, 24–26, 56, 91
  Specialists, role of, 25
Manufacturing Advisory Service (MAS), 79
Manufacturing, Organisation, People and Systems (MOPS), 192
Mapping, 47–53
  Decision Point Analysis, 49
  Forrester Effect, 49
  Overall Structure Map, 49
  Process Activity Mapping, 48
  Product Variety Funnel, 49
  Quality Filter, 49
  Supply Chain Responsiveness, 49
McCarthy, Dennis, 11, 31, 141, 143, 156
Medical Devices Company, 6, 119, 167, 194
Mentoring Companies, 3, 4
Monden, Yasuhiro, 53, 125, 127, 130, 140, 182
Motorola Inc., 101, 110

Nakajima, Seiichi, 68, 141, 145, 146
National Vocational Qualifications (NVQ), 205
Networking, Industrial, 190

Ohno, Taiichi, 16, 48, 59, 124, 125
Organisational Learning, 185, 191
Overall Equipment Effectiveness (OEE), 68, 87, 145, 147, 149, 155, 156, 160, 195

Participating Companies, LEARN 2, 5–7
  Air Repair Company, 6, 118, 194, 196, 198, 199
  Cosmetic Company, 5, 72, 89
  Health Products Company, 5–6, 119, 195, 197, 198
  Heavy Products Company, 5, 194, 195, 197
  Mornington Cereals Company, 6–7, 195, 196, 198
Parts Per Millions (PPM), 11, 38, 110
Policy Deployment, 54–57, 58, 92, 168
Problem Solving, 97–99
  Cause and Effect Chart (CE Chart), 88, 98–103
  Plan, Do, Check, Act (PDCA), 67, 100, 110, 168, 170, 182
Process Improvement, 164, 165, 166, 170, 179
Pull Production, 15, 125, 126, 138
  Kanban, 54, 90, 125, 127, 128, 129, 131–134, 139
Push Production, 124
  Materials Requirements Planning (MRP), 124
  Manufacturing Resources Planning (MRPII), 124

Quality Assurance, 105, 107–108
Quality Management Stages, 105

Return on Capital Employed (ROCE), 96, 110
Return on Investment (ROI), 96, 110
Rich, Nick, 3, 11, 14, 27, 31, 48, 54, 55, 60, 84, 87, 88, 96, 141, 143, 144, 147, 155, 156

Sekine, Kenichi, 130
Senge, Peter, 59, 190
Seven Wastes, 16, 17, 48
Shainin, Dorian, 107, 111
Shingo, Shigeo, 16, 140
Shirose, Kunio, 79
Single Point Lessons, 157
Six Sigma, 38, 95, 96, 99, 101, 105, 110, 112, 119, 121, 147
Society of Motor Manufacturers and Traders (SMMT) Industry Forum (IF), 163, 169, 180
Suppliers, 41, 42, 202, 204
Sustainability, of Improvements, 163, 167, 171, 173–176, 180
  Sustainability, Class A, 166, 171
  Sustainability, Class B, 166
  Sustainability, Class C, 166
  Sustainability, Class D, 166
  Sustainability, Class E, 166
Systems Perspective, 7

Takt Time, 50, 130, 131
Technology Transfer, 190
The Machine That Changed The World, 1, 2, 3, 14, 31
Theory of Constraints (TOC), 112
Total Productive Maintenance (TPM), 9, 87, 141, 146, 147–149, 161–162, 174, 178
TPM, Pillars of, 151–152
  Office TPM, 161
Total Quality Management (TQM), 9, 16, 54, 58, 84, 93, 95, 99, 108, 109, 110, 112, 147
Toyota Motor Corporation, 1, 9, 13, 14, 16, 39, 125, 126, 130, 141, 202, 204
Toyota Production System (TPS), 9, 59, 122, 124, 125, 139, 140
Trade Unions, 21, 40, 77, 194
Traditional Organisations, 33–36

Value adding, 28
Value Stream, 41, 48, 50, 164, 167, 182
Variation, 111
Visual Management, 57, 80–94

Willmott, Peter, 141, 147
Womack, Jim, 11, 31, 105, 110, 122, 125, 143
World Class Manufacturing, 13, 53, 62